P9-EMF-027

Observing Earth's Lifestyle

Observing Earth's Lifestyle

Considering the Value of Earth Stewardship
in Environmental Protection

David L. Romain

Copyright © 2017 David L. Romain

All rights reserved.

ISBN-13: 9781519678515
ISBN-10: 1519678517

Preface

WHEN I STARTED this study, I was focused on a geographic/ecological analysis of Earth. It was directed toward the challenges of the ongoing climate disruption. But, the combination of my geographic, community economic development, teaching, parent involvement, urban planning experiences and applied research backgrounds would not let me reduce my awareness to any single specialty.

Most of all my accumulated awareness of the environment rebelled against the modern propensity for focused specialization and isolated individualism. I have seen the continuous interdependence of human life and all other life on the planet. The modern lifestyle is so focused on the individual, the spirit of interdependence seems like a distant after-thought in the general scheme of things. Yet, Earth's lifestyle flows with expressions of interdependence at every juncture. It is the ultimate power that drives all life on the planet.

The modern lifestyle has not taught us to live in harmony with Earth's lifestyle. In fact, we do not see the value of Earth's nature. We have not learned to value the planet's resources in the context of Earth's desire to serve all of life equally. Thus, we tend to see resources merely as commodities and other life on the planet as either isolated conveniences for our pleasure, or nuisances to be avoided. Each commodity, which we value, is seen as an isolated object for convenient individual consumption.

From such a vantage point, it has become extremely difficult for us to appreciate the personal level at which a solution to this current climate disorder requires us to transform our lifestyle for our own survival as interdependent members of the global human community.

By their very nature, governance systems pose a communal expectation of mutual care, based on their implied promise to serve the community. The modern lifestyle is based on the colonial system, which has built today's global economy. Modern governance systems build community and hold those communities together in order to support the global economy. This global environmental crisis underlines the need for governments to represent the common good without compromise.

The global economy represents an advanced state of interdependence. It is a matrix of exchanges that affects and depends on the capacity of communities the world over to exchange goods and services. That is global interdependence.

Responsible governments owe it to their respective communities to protect the environment on behalf of its citizens. When the governor of one state declares the efforts of his community to take steps to ease the crisis are outlawed the democracy of that community has broken down. This is why the actual function of ongoing democracy is critical to the survival of humanity. Functioning democracy is a requirement, not an option in the implied quest for human survival.

At best, we are all tenants on Earth. Each of us come into the world carrying nothing. Each of us is a spiritual being, occupying a physical body. That body has a life expectancy and, when it expires, our spiritual being is left with nothing physical. This has been the firm understanding of the world's traditional societies, which have dominated human history until recently. That is why it had been so natural for these societies to seek to live in harmony with nature.

Nature operates in the spirit and focus of Earth's lifestyle. Those, who live in the spirit of gratitude for the freedom of living on Earth naturally adopt that lifestyle in appropriate humility. In so doing, they learn to live in harmony with all of Earth's natural events, such as earthquakes. volcanoes, hurricanes, forest fires, floods and the continuous patterns of weather and climate. Accordingly, these events seldom became disasters for them.

Built into their sense of freedom was the context of mutual accountability. For them freedom was permanently attached to responsibility. Freedom was not the free-for-all that modern society has come to understand and expect.

The traditional outlook to life encouraged them to manage their violence in ways that allowed for people to heal each other's pain and suffering with elaborate feasts of reconciliation, whenever they needed to come back together after a feud. Their violence was a closed circle, which anticipated and expected a reconciling closure.

Acknowledgements

THIS IS THE culmination of several years of study, observation, curious questioning, general interest, obsession, or mere passing fancy throughout my life. Along the way, I have been influenced by friends and family, as well as enemies, retractors, special events and a host of external stimuli. However, through it all, my deepest gratitude goes to my wife, Kathy, who would not allow me to quit the task at hand. My sons, Tim and Andrew were always very supportive of my ideas. Yet, they kept me fully aware of the next generation, with a sense that somehow their generation and that of their children were living in a world severely compromised by the irresponsibility of my generation

Teaching physical geography to community college students prompted me to teach about the emerging phenomenon of global warming as we entered the new millennium. The issue oriented US public, without a familiarity with basic geography left my students thoroughly confused by the landscape of global warming opinion. Thus, considerable gratitude goes to my many students over a span of several years, who kept asking the questions that drove me to expand my research into what I have since compiled.

In order to address many of my students' questions and concerns about the real threat of global warming, I relied heavily on friends and colleagues with expertize in the wider field of environmental awareness.

Rita Haberlin, my first faculty advisor in the Peralta Community College District helped me organize my course material, from where the idea to write about global warming was born. Carl Anthony, of the Ford Foundation and Earth Island Institute shared generously his ideas and concepts from Earth Island Institute; Gar Smith of the Institute also shared his particular expertise on nuclear energy. Colleagues in the geography departments of the colleges of the Peralta Community College District, in Oakland, California were similarly encouraging.

Teaching at the community college level had many unexpected opportunities to learn. From retired experts to students with unusual backgrounds, I learned unique concepts to enhance my understanding of Geography. A retired airline pilot gave me perspectives on the atmosphere I would not have had otherwise. One student was an installer of solar panels, who could explain to the class how the solar panels worked to produce electricity. Some class term papers challenged me to do additional research. One of my students, Javier was the son of a refugee from El Salvador, who had visited his father's homeland regularly. He took me to meet his father, Julio Leiva, who wrote one of the books in my bibliography: *Un Mejor Manana*, El Terco Deseo de Crear, with minute details of the war in El Salvador and the refugees it spawned.

Teaching school could be limited to a one-way experience. However, my deep friendship with the late Clarence Jackson, an education psychologist and advisor the California State Department of Education transformed my mere teaching activity into an active ongoing teaching-learning process. Without that exposure, I would not have learned so much from my effort at teaching community college physical geography.

Eventually, I travelled to Belize on a vacation and found it to be an ideal environment conducive to writing about earth stewardship. The place offered me enhanced familiarity with tropical forests similar to the forests of my youth in Trinidad and I found it to be a very supportive community in general. The deeper impact of climate disruption and colonialism is much more detectable here.

While in San Ignacio, I became acquainted with the Creation Care Study Program, Belize, an academic program, which offered study for college credit. The program emphasized sustainability as a composite part of environmental stewardship and observed local land use and activities, which demonstrated this ideal. I am especially grateful to Micalagh and Josh Moritz, who invited me to accompany their students on field trips and get to see traditional farming and land cultivation that honors ongoing stewardship.

My final assessment of this project would be amiss if I did not extend special gratitude to Richard and Sharron Singleton, Justin Cannon, Betty Lane, Margaret Young and Marilyn Wall, as well as the folks at Adelynrood Conference Center, in Byfield, Mass., who invited me to conduct workshops on the topic of Earth stewardship.

Dedicated to
Tiffany Arianna Grace, my first granddaughter
and all, who would share planet Earth with her,
with the fond hope that it would become more
human-friendly habitat than what it is today.

Table of Contents

Introduction

EARTH IS ALIVE! The planet is an exciting place, full of mystery. Earth's life gives us life. Powered by the sun, which holds this planet in orbit, Earth revolves around the sun. Everything we need for living comes from Earth,[1]. It is made available to all life on the planet and is freely given. Under normal conditions, Earth provides its resources in abundance. Scarcities exist only where our rate of consumption exceeds Earth's rate of replenishment.

The more I learn about it, the more I come to appreciate how much I still do not, (and never would), know about the planet. Thus, I still approach Earth with far more questions than answers. And, many of the "answers" are merely educated hypotheses, awaiting further challenge and clarification.

In elementary school, on my home-island of Trinidad, in the Caribbean, I was first introduced to the world as a place larger than the place I would relate to on my way to school. I remember the first time I was shown a globe and the tiny speck that represented Trinidad. I could look out of my window at home and see half of the island. How could the world be so large?

I was told that, in time, I would get to know a great deal about this huge globe on which I lived. The places were different, as were the people who lived there.

I had observed that first-hand, when a family friend from Aruba in the Dutch Antilles visited, she would speak to her children in Dutch, to my father in a French Patois and to me in English. Her oldest son also spoke English and I would have the opportunity to reconnect with him over fifty years later in Boston, where he managed the city's largest community development corporation.

Still, the wide range of languages continues to represent the abundant array of people, communities and cultures that make up our global village. The oat meal package on my kitchen pantry shelf, above all, intrigued me as a child. It had directions in eight languages, including Arabic and Chinese. To me, the planet was and is still shrouded in overall mystery, more of which I continue to explore.

From what we understand of the existence of Earth today, if its life span of four billion years were scaled into one calendar year, human existence would have begun within the last few minutes of December 31st. So, what could we "know"? What I do know is that everything we need and use for our daily living comes from the planet. Each of us is born on the planet with nothing and when our bodies expire, each in turn leaves with nothing. Thus, I am encouraged to echo the observation of Chief Seattle: "I belong to Earth. Earth cannot belong to me!"

Earth is our most precious gift. The gift comes with everything on it, free for me to partake of. As such, I am eternally grateful. That gratitude has led me to pursue a concept of "Earth Stewardship", in which I seek to act out in appropriate response to the Gift of Earth.

I was born into and have lived in this tradition throughout most of my life. Through this tradition, I gravitate toward a concept that Earth presents itself to us as a free gift, whose bounty we should share equally with all life. In order to share with all life on the planet, we must first take the time to learn how all life on the planet is fully integrated as an interdependent collection of individual parts.

As such, I feel we are called to care for the planet. That caring is my concept of Earth stewardship, through which I seek to honor Earth. On Earth we are all mutually interdependent parts of an eternal life cycle. Life is a gift! An appropriate response to such a generous gift of life should be GRATITUDE.

Stewardship is a natural way to participate in Earth's life cycle. The animals seem to play their part in taking care of the resources important to their daily livelihood. Those that hunt, generally eat what they kill. What they don't eat almost always is eaten by other scavenging animals. Trees grow to hold the soil together, so that it can store the living water, without which we could not live. In places far from the ocean, the trees transmit soil moisture into the atmosphere to create clouds, which eventually produce rain. Some of the rain, along with melted snow finds its way into rivers, which flow to the sea. The rivers carry sediment to the ocean, which becomes compacted into sedimentary rock.

In traditional society throughout human history, people have patterned their lives after the dictates of their environment, much like the animals, birds and plants in their habitat. They sought to live in harmony with all life, which also shared Earth. The planet's harmony incorporates four interrelated cycles. However, the energy driving Earth's harmony radiates from the sun.

The Sun holds Earth in its orbit as it harmonizes the relationship between the following four distinct cycles, [2]. The Atmospheric Cycle is an envelope of air, (the atmosphere), which surrounds the planet. It contains mostly fixed amounts of nitrogen, oxygen, argon, carbon dioxide and varying quantities of water vapor. Animals breathe oxygen, while the plants consume nitrogen and carbon dioxide.

The Atmospheric Cycle propels the Hydrologic Cycle. Water carried by clouds in the atmosphere can be transported to different parts of Earth. The ocean water evaporates into water vapor, which condenses into clouds, which fall as rain, some of which runs off into rivers that flow back to the sea.

The rivers connect the Hydrologic Cycle to the Rock Cycle, because the water that flows in the rivers provides the greatest form of rock movement on earth. As the water flows over the rocks, it breaks them down into eroded sediment. The flowing water transports the sediments, grinds them into progressively smaller particles and smoothing them in the process. The river eventually deposits sediment in the sea, where its flow is terminated. The sediment is made up of all types of rock material. It is compacted to form sedimentary rock.

At lower depths in Earth's crust and into the mantle, heat is intensified sufficiently to transform some sedimentary rock into metamorphic rock. Where all these rock types are close to a pool of magma, the excessive heat of the magma melts adjacent rock which is added to that magma pool. When magma is cooled it may eventually solidify to form igneous rock. This transforming of rock types, one to another, is the essence of the Rock Cycle.

Magma, whether it is near Earth's crust, or deep in the mantle, is essentially the product of the movement of tectonic plates in the Tectonic Cycle. The planet's crust is made up of twelve tectonic plates, which roughly overlap each other, leaving no space between them. These plates continually jostle each other, as some diverge from each other while others converge on each other.

The Discussion Format:

I live in California, USA, and recently retired from teaching at various community colleges. So, I naturally speak to a California and US audience. Much of what I discuss, however, is applicable to a global consciousness. It is particularly applicable to the global consciousness that identifies its lifestyle as modern.

In their own ways, the traditional societies of the world have lived by a wisdom that honors the seasons and customs of the Earth beneath them. I call it "Earth's lifestyle". They have considered it to be the standard from which they patterned their own lives. Central to this wisdom is a genuine appreciation for the priceless and unconditional gift of "Mother Earth".

Most traditional societies seem to have lived by a common spiritual code of honor. Their spiritual practices may have been unique. However, these practices observed the cycles, rhythms, seasons and general patterns of earth's way of presenting itself. Earth provides, equally and unconditionally, for **ALL** the life on the planet.

This book is presented in distinct sections, in an attempt to systematically conceptualize Earth's lifestyle:

<u>Section I</u> describes the essential concept of Earth's lifestyle. In contrast, our modern lifestyle does not acknowledge Earth as the center of its existence. We live as if Earth is a disposable resource, merely a convenient landscape on which we "hang out". Our primary goal is to exploit the planet for our own individual and personal convenience.

<u>Section II</u> explores stewardship as a commitment by us to each other and to Earth. In other words, stewardship helps us to become familiar with Earth's lifestyle. Such knowledge may guide us to best benefit from what Earth has prepared for life on the planet. By developing a caring relationship with Earth based on gratitude our effort to follow Earth's lifestyle would become more natural.

Earth's lifestyle is presented to us daily and continuously, from dawn to dusk and back to dawn. Our awareness of its lifestyle is accentuated in events, such as: earthquakes, volcanoes, hurricanes, floods, tornadoes, forest fires, tsunamis, avalanches, etc. We study the anatomy of seven of the major natural events in order to avoid experiencing them as disasters in the future, as we have so often in the recent past.

As we focus our awareness of Earth's lifestyle we should become more attentive to these natural events. By examining the ways in which traditional peoples adapted their lifestyle to fit Earth's customs we may see where we could develop policies and customs more effective to accommodate and be in tune with ongoing natural events. Accordingly, we explore ways to adapt current land use practices to ensure safe human habitation in vulnerable areas.

<u>Section III</u> introduces the concept of a sustainable village and its potential adaptability to Earth's lifestyle. Stewards of Earth can help us to appreciate how our negative attitude toward Earth is often at the roots of many disasters we experience, both "natural" and industrial.

Human existence on the planet has been about one million years. Until some 5,000 years ago, virtually all life on the planet existed in villages and small settlements seldom more than a few thousand people. As such, most people lived in villages a radius of a few dozen miles and were acquainted with adjacent villages, with whom they traded. They seldom ventured beyond the neighboring villages, for fear that they may wander into the territory of other people unknown to them, [3].

Section IV looks closely at the planet's resources, with particular emphasis on those resources that drive the global economy. In this respect, we explore those customs that have encouraged us to abuse the natural market economy of free exchange by instituting an unfair, deceptive market system. This modern market system has redistributed Earth's resources, (and related wealth), into the hands of a few, creating an enormous volume of global poverty.

The modern market system has also willfully abused Earth's resources, creating a critical global climate crisis, which threatens to terminally establish a new environment hostile to future human existence. This attitude of abuse has been extended to our ongoing care of the very infrastructure that sustains our urban industrial lifestyle. We discuss how our attitude of negligence to our urban infrastructure is similar to our neglect of Earth, thus, ultimately self-destructive.

Section V explores the extent to which the excessive consumption of a privileged few has impoverished many, supported by the widespread deceit of a ruling class. In the meanwhile, Earth's lifestyle prevails in its customary system of sustainability. Stewardship is a concept, which values sustainability in the use of natural resources.

Stewardship encompasses a sense of genuine caring for *all of life* on the planet. This means that all resources are meant to be shared more equitably throughout the global community. Without more equitable sharing future human survival is increasingly unlikely in the long-term.

Basic stewardship essentially requires a sense of mutual respect for all life, in the context of "live and let live". As such, the book emphasizes the need to be grateful for Earth generous gift of abundance. We have to allow Earth to produce its natural abundance, from flowering plants and weeds to reproduction of most animals, especially insects, fish and many aquatic organisms.

Much of the transformation necessary to live sustainably and in harmony with the planet, is already being piloted and demonstrated in many parts of the world. Paul Hawken, Andres Edwards and many other researchers illustrate examples of many efforts at practical stewardship that have proven currently to be successful throughout the world, [4].

These examples together demonstrate an emerging global practice of stewardship. The challenge is urgent. However, it is already receiving an significant public response. When individualism gives way to global cooperation in this effort, more widespread stewardship will follow.

The ultimate reality is that the present popular lifestyle is fundamentally unsustainable. Earth will survive us, along with many other organisms that live independent of the resources critical to human survival. As such, we are systematically destroying our human resource survival base.

Our arrogance will not destroy Earth, because, we merely occupy a small part of the planet, at its surface. The physical Earth shall survive our self-destruction.

Notes

1. This document will continue to capitalize Earth as if it were a proper name, in order to personalize Earth and to keep in focus the idea that planet Earth is alive. My premise is that Earth is alive and offers a continuous example of life, as it should be lived. Earth is in control of all natural events. I refer to the overall pattern represented by this control as Earth's lifestyle. I believe Earth invites us, by example, to live in harmony with this lifestyle.

2. Mr. Christopherson discusses the Geologic Cycle as a system combining the Hydrologic Cycle, the Rock Cycle and the Tectonic Cycle. The

Atmospheric Cycle had been added, distinct and separate from the Hydrologic, since each has a unique place in the operations of Earth's Cycles. See: Robert W. Christopherson: Geosystems, seventh edition, c2009, Pearson Prentice Hall, Figure 11.5: "The Geologic Cycle", p. 332.

3. Jared Diamond, in his book: *The World until Yesterday*, describes in great detail the life styles and customs of people in traditional societies throughout the world, many customs, which are still in practice today. See: Diamond, Jared, *The World until Yesterday*: What can we learn from Traditional Societies? Viking Press, New York, c2012

4. The following two authors present separate, but complementary ideas, upon which this book relies in promoting the concept of Earth stewardship: Andres R. Edwards: The Sustainability Revolution, Portrait of a Paradigm Shift, New Society Publishers, c2005 and Paul Hawken: Blessed Unrest, How the Largest Movement In the World Came Into Being *and* Why No One Saw it Coming, Viking Press New York, NY., c2007.

I

Four Cycles and a Sun

Earth's Lifestyle in Perspective

A STUDY OF Earth's geography gives us substantial evidence of the intricate handiwork in its basic structure. Thus, we can witness the dramatic interdependence of the sun and four dynamic cycles, (atmospheric cycle, hydrologic cycle, rock cycle and tectonic cycle,) that work together to drive Earth's ongoing function.

Earth is a living planet that speaks to us all, through its lifestyle. As we observe the habits, customs, patterns, systems and cycles of Earth, we become more familiar with "the way Earth works". Furthermore, if we could realize how this planet was a few hundred years ago, we may grow to appreciate what had been left to us by distant ancestors and the natives of many indigenous lands.

In the cycle of Earth's resources, nothing is wasted. Whatever organism lives, eventually dies, leaving its expired body to be eaten by other organisms or to decompose, feeding still other living organisms. In this way the life-force of Earth continues and each of us borrows from that life-force to keep our temporary bodies alive for as long as possible.

The entire scope of Earth's natural operations is patterned after endless, sustainable cycles. From the cycle of each day to the seasons of the year, or from the panting of seeds to the harvesting of the products of those planted seeds, these cycles continue to support us and each other through a matrix of self-sustaining activities.

The Miracle of the Four Cycles:

Everything that happens physically on Earth is related to the function of four distinct cycles and their relationship to the sun. Each cycle may relate differently to the sun. However, each, in its own way, has been influenced by the sun over billions of years.

Atmospheric Cycle: As an envelope of air, the atmosphere surrounds the planet. It contains mostly fixed amounts of nitrogen, oxygen, argon, carbon dioxide and varying quantities of water vapor. Animals breathe oxygen, while the plants consume nitrogen and carbon dioxide.

Earth's surface is heated unevenly by the sun, which, in turn, heats the air above it. An unevenly heated environment, causes constant variation in atmospheric pressure. Winds blow from high pressure areas to low pressure areas. In the process, the winds redistribute heat and cold respectively, around the planet.

Another important miracle performed by the atmosphere is its unique ability to absorb water. The atmosphere discriminately extracts pure fresh water from the oceans on Earth, which constitutes over 97% of all Earth's water. In this way, the Atmospheric Cycle propels the hydrologic cycle.

Hydrologic Cycle: Most of the water vapor found in the atmosphere is being extracted from the salty ocean, as pure fresh water, through evaporation. The drier the air, the more water it will absorb. For instance, the dry air in your room in winter dries out our skin, as it extracts the body's water. After all, our bodies are over 80% water.

Water carried by the atmosphere can be transported to different parts of Earth. The ocean water evaporates into water vapor, which condenses into clouds, as they travel across the sky. The clouds bring rain, some of which runs off into rivers that flow back to the sea.

When it snows, some of that snow eventually melts. Much of the fallen rain and melted snow seeps into the ground to become ground water. Some of that water eventually seeps back into rivers to keep them flowing long after the rain has stopped. Trees extract much of the ground water through their roots, carrying nutrients from the soil for their own food. The excess water they take in is "transpired" by the leaves back into the atmosphere as water vapor to contribute later to more cloud formation, which once more ride the passing winds.

The Rock Cycle is driven by the rivers of the Hydrologic Cycle, as the flowing water provides the greatest form of rock movement on earth. As the water flows over the rocks, it breaks them down into eroded sediment. The flowing water transports the sediments, grinds them into progressively smaller particles and smoothing the standing rocks in the river bed in the process. The river eventually deposits its sediment in the sea.

On the edge of continents, the "continental shelf" consists of sediment that has been compacted over millions of years to form sedimentary rocket its lowest layers. Within the planet's body, the sedimentary, metamorphic and igneous rock types are continually being transformed into each other in a never-ending cycle.

The intense heat of the planet's interior melts together the different types of rock to form magma, which eventually feeds volcanoes. It comes to the surface as lava, which solidifies to form igneous rock. All three rock types are mixed together and deposited as sediment in the sea, where it is scattered by the tides. The sheer weight of all the settled sediment causes the upper layers to compact the lower layers into reconstituted sedimentary rock, as it builds up.

At lower depths in Earth's crust, heat is intensified sufficiently to transform some sedimentary rock into metamorphic rock. Where all these rock types are close to a pool of magma, the excessive heat of the magma melts adjacent rock, all of which is eventually cooled to form igneous rock all over again. This is the essence of the Rock Cycle.

Tectonic Cycle: Magma, whether it is near Earth's crust, deep in the mantle, can also be the product of the movement of tectonic plates in the tectonic cycle. The planet's crust is made up of twelve tectonic plates, which roughly overlap each other, leaving no space between them.

These plates continually jostle each other, as some diverge from each other while others converge on each other. The diverging plates release magma, which instantly fills the space created by the divergence. As the newly released magma solidifies upon cooling at Earth's surface it forms new crust, in the form of igneous rock.

The converging plates generally cause the heavier plate to "subduct", (submerge forcefully), beneath the lighter plate with considerable friction. The friction generates a high concentration of heat, sufficient to melt the rock through which the subducting plate travels, creating pools of magma. This magma works its way upward, often to emerge at Earth's surface as a volcano. The Tectonic Cycle provides much of the magma necessary to start the Rock Cycle in the crust and upper mantle.

The Sun: All of Earth's activities are in some way related to at least one of these four cycles. Earth itself is held in orbit by the sun's gravity.

Earth revolves around the sun, in continuous motion. Each complete revolution around the sun takes approximately 365.25 days. In addition, each day is marked by Earth's rotation on its own axis. Each a complete rotation takes 24 hours.

The sun's other role in Earth's life provides most of the planet's surface heat energy. The heat is unevenly provided and distributed and this uneven distribution enables the function of the Atmospheric Cycle.

Earth is sufficiently endowed to feed, clothe and generally satisfy the basic needs of all life on the planet. So complete is the provision of Sun and Earth, that they are both self-cleaning and self-healing. There is nothing we can do to scar Earth, which is not self-healed in its own time. We only inflict "flesh-wounds"!

CHAPTER 1

Here Comes the Sun

IT SEEMS THAT the sun was there first. Earth came later to revolve in orbit around our daystar, the sun. However, from our vantage point, after the long dark night, on each new day "here comes the sun".

The sun's radiant heat provides most of the surface warmth we feel above ground in our respective habitats. It also provides our daylight, a welcome sight to those who live sufficiently far from urban centers, where night lights cancel out the complete darkness of truly rural nightlife.

The sun provides the heat that generates many of the weather features of the atmospheric cycle. Because the planet is an orb, the part that is in the most direct light rays from the sun, (where it is at its highest point in the sky), receives the greatest intensity of the sun's heat energy. The farther one is from the direct rays, the less the heat intensity.

After it heats Earth's surface, that heat is redistributed in many ways. Some is absorbed by the land, (the body of the planet), while much of the rest is reflected back into the atmosphere to determine our ambient temperatures, [1]. Sunlight also provides plants with the nutrients they need to produce chlorophyll for the building of proteins.

The sun has another influence on Earth. It holds Earth in a specific spatial relationship to itself, through the force of gravity. As such, Earth travels in an elliptical orbit around the sun, in a permanent state of rotation on its axis.

At all times, half of Earth is in the sunlight, while the other half is in shadow. The sunlight gives us daylight, while the shadow brings us night. Thus, as it rotates, every spot on Earth, between the Arctic and Antarctic Circles, rotates into daylight and back into darkness. Clouds may block our view of the sun by day and block the moon by night. Both the sun and the moon shine behind the clouds on those cloudy days and nights.

In fact, it is not unusual for a cloudy day to significantly change the temperature and humidity from what it had been without the clouds. The resulting overcast can also serve to hold the ground temperature more constant than the cloudless alternative.

Earth's Rotation:

Earth makes one complete rotation every 24 hours. Earth's rotation can be used to measure time of day. We use this phenomenon to measure time zones and the International Dateline. The lines on a globe used to measure the rising of the sun around the world are called longitude lines. They stretch from the North Pole to the South Pole and are measured in degree angles from a Prime Meridian of 0 degrees longitude, which passes through a point in Greenwich, just east of London, England. Accordingly, when it is noon at 0 degrees, it is after noon east of Greenwich and before noon west of Greenwich.

Since a circle has 360 degrees, Earth rotates through 360 degrees every 24 hours, (single day). By dividing 360 degrees by 24 hours we can use a 15 degrees rotation to represent each hour. When it is noon in Greenwich, it is 7am in New York at 75 degrees West Longitude and 2 pm in Durban, South Africa at 30 degrees East Longitude.

The sunlight makes the daytime warmer and the absence of sunlight allows nights to get cooler. From the sub-tropical regions to the poles, summer days are significantly longer than 12 hours and during the winters, days are significantly shorter than 12 hours.

Moreover, as we approach the Polar Regions in the north and south, the length of a summer day increases toward at the Arctic Circle and the Antarctic Circle, where the longest days are 24 hours of sunlight and the shortest days in the winter are 24 hours of darkness, (or moonlight!). Where the days are longer, daytime average temperatures are higher than the days with shorter daylight hours.

Earth's Revolution:

The Earth rotates on its axis at a perpetual 23.5 degree angle from the vertical. Because of this angle, Earth's revolution around the sun results in the North Pole being tilted toward the sun in the summer and away from the sun to give us our winters. Earth revolves around the sun once every 365 ¼ days. The North Pole is tilted furthest toward the sun on June 21st. The tilt away from the sun is greatest on December 21st.

In the southern hemisphere, the seasons are reversed, because, when the North Pole is tilted toward the sun, the South Pole is tilted away from the sun. When it is summer in the north, as a result, it is winter in the south.

In most of the tropics, people see the sun rise between 5:15 and 6:45 am each morning and set between 6:45 and 5:15 pm each evening. In this respect, the length of daylight varies little throughout the year. The north may see the longest day approaching 14 hours around the latitude of the Tropic of Cancer, when the sun is overhead. The shortest day never less than 10 hours and that occurs when the sun is overhead at the Tropic of Capricorn. At the Equator, the days vary little from 12 hours year round.

Another unique feature of the tropics is that each location within this region experiences two "longest days". This happens when the sun is directly overhead at that location, as Earth's revolves around the sun, causing the sun to seem to travel between the Tropics.

As one travels north from the Tropic of Cancer, the summer days grow longer and the winter days get shorter, eventually we see 24 hour of daylight within the Arctic Circle during the summer and 24 hours of darkness in the winter, [2].

Seasons:

As Earth revolves around the sun, the path of direct overhead sunshine travels between the Tropic of Cancer and the Tropic of Capricorn in a continuous flow. On June 21st the sun is overhead at the Tropic of Cancer on its way to the Equator, where it arrives around September 21st. The overhead sun continues to the Tropic of Capricorn, arriving around December 21st, from where it returns to the Equator by March 21st on its way back to the Tropic of Cancer again. March 21 and September 21 dates are each referred to as the Equinox and the June and December dates are each called the Solstice.

Seasons in different parts of the globe are influenced by the angle of the noonday sun. It gets lower on the horizon as we go from the Equator toward the poles. The intensity of the sunlight decreases, as the noonday sun gets lower on the horizon. Sunlight produces a decreasing amount of heat in places closer to the poles; hence the poles are colder than the tropics. At the poles, the sun gives light, but its heat is negligible.

Close to half of Earth's surface is in the tropics. This zone experiences overhead sun at least twice annually, as the directly overhead sun drifts north and south of the equator. The sun in the tropics is always close to an overhead angle, causing the temperature to be hot during the day and warm at night at sea level.

Seasons in this half of the world are influenced purely by rainfall. The tropics experience a dry season and a rainy season, depending on the relative amounts of rain received. Some places actually experience a separate and distinct period of drought in the dry season. In contrast, the rest of the world experiences four distinct seasons: spring, summer, autumn and winter.

Within the Arctic and Antarctic circles, the temperature sometimes can change dramatically with a shift in the winds. Here, the prevailing winds are the polar easterlies, which blow from the high pressure center, located at the respective poles. However, the greatest shift takes place with the steady change in the length of daylight during the year.

In contrast, locations in the mid-latitudes see much greater variations in weather patterns from summer to winter. The mid-latitudes, located as they are between the extremes of the polar region and the tropics, are influenced by both. In this region, the prevailing winds are the westerlies, which push air masses from west to east. Within these air masses a low pressure center develops.

Each air mass creates a weather systems of its own, with winds being blown in toward the center. These are cyclones, with a lower atmospheric pressure at their center is not much than the pressure at its outskirts. The resulting winds are mild. When the low pressure centers increases in intensity, relative to the outer air mass, the cyclonic winds are stronger and can draw in large quantities of both warm and cold air. The temperature mix may often create violent weather, such as super-cell thunderstorms and tornadoes, especially in the spring and fall seasons.

As these air masses are pushed eastward by the prevailing westerlies, they are influenced by adjacent weather conditions. The air masses are generally about 1,000 miles across. Thus, if they travel close to the tropics, they tend to draw in warmer air and bring a welcoming break in the weather during the winter months. Where they travel close to the polar region, they tend to draw in polar air, creating or facilitating the periodic "cold air blasts" during the winter months and can even transport freezing temperatures to the adjacent sub-tropical regions, such as Florida, USA in North America.

In the tropics, the prevailing winds blow toward the equator from both the north and the south. These winds are more dominant than the westerlies of the mid-latitudes. They are the Northeast Trade Winds and the Southeast Trade Winds. They both blow toward the Equator form the northeast and southeast respectively. These winds originate in a region a few degrees pole-ward from the Tropic of Cancer and the Tropic of Capricorn.

The Circle of Illumination:

At any time, one-half of Earth is in sunlight, while the other half is in darkness. The line which separates the light from the darkness is called the Circle of Illumination. The Circle is fixed, in relation to the sun.

However, since Earth is in continuous rotation, this boundary between day and night shifts steadily like a moving shadow as the planet rotates daily and as Earth travels in its orbit around the sun.

When the North Pole is tilted furthest toward the sun, the Circle of Illumination completely contains the Arctic Circle. As the planet rotates, everything within the Arctic remains in the sunlight, giving it 24-hour days. This event occurs on June 21st. At the same time, the Circle of Illumination excludes the Antarctic Circle, leaving it in 24 hours of darkness, during winter in the Southern Hemisphere.

Time and the Weather:

We in the north tend to regard June 21st as the beginning of summer and December 21st as the beginning of winter. Summer does not actually start on 21st June, nor does one winter always wait for 21st December to begin. Summer does not begin on the same day in Georgia as it does in Minnesota in any given year. Also, at any location in different years, summer or winter follows a different time frame.

Moreover, because other factors go into creating our weather, actual daily temperatures and seasons are seldom duplicated on any date. The weather is created by whatever is happening at the current time. Climate summarizes general weather patterns over a period of several years.

Thus, when you listen to your local weather report it almost always reports that today to be above or below the "expected" temperature for that date. That "expected temperature" is actually a mathematical average of at least 20 years of recorded temperatures. Accordingly, the actual temperature for that day in question can be anywhere in the mathematical range of temperatures recorded in the 20-year range recorded earlier, especially in mid-latitude locations.

For those who live in the mid-latitudes, (between 30 and 60 degrees Latitude, north and south of the equator,) they often see huge changes in the weather almost every few days, especially during the winter. It *is* normal to have the freezing weather interrupted by mild spring-like days. Similarly, summers sometimes include unusually cool days.

The Sun's Influence on Climate:

If longitude lines represent time of day, latitude lines indicate temperature. As we move away from the equator, daily and seasonal temperatures tend to decrease. As the angle of sunshine striking the land determines the land's temperature, the Polar Regions are significantly colder than the tropics.

The sun provides continuous radiant heat to Earth. The planet is made up of land and water. About 72% of Earth's surface is under water. Both land and water are enveloped by the atmosphere. Heat energy from the sun is captured by both land and water. The two media absorb and redistribute heat differently. The land allows the heat energy to travel through its body in a process called conduction. In contrast, the water, (and air), distribute heat through the process of convection. Air allows heat energy to travel more quickly than both land and water. Land heats and cools more quickly than water.

Surface water is heated by a process called convection. This process causes water to be heated at the surface and be mixed with water slightly beneath the surface which has not yet been heated. This mixing is a slow process, since warmer water tends to rise above the cooler water. As the molecules of warmer water heat up, they expand and set in motion small water currents. These currents begin to mix warmer molecules with colder molecules, warming the larger body of water. These are called convection currents. They are similar to the currents of air responsible for warming and cooling the ambient temperature around us daily.

The net effect of heat energy in our midst is that land exposed to the same heat energy as water will be heated faster than water. Land also responds to cooling influences faster than water. Air is able to heat and cool both land and water.

However, because air is so quickly influenced, it is also influenced by the land or water over which it travels. Accordingly, in winter a cold body of air may hit Minnesota with temperatures of -20 degrees C and cool the entire Mississippi Valley. When it finally arrives in New Orleans it may have itself warmed up to +5 degrees C.

Most of the heat we feel in the atmosphere comes from the sun, but not directly. The heat we feel is reflected from the heat stored on Earth's surface, land and water. The atmosphere has the distinct advantage of being able to absorb and release heat most rapidly. That is why the wind, (air in motion), is so significant in bringing us weather changes, especially when is blows from water onto land.

The length of day influences temperature, such that long days facilitate warmer temperatures in the summer. Short days provide lower temperatures in winter, [3].

The Interactions among the Four Cycles:

As suggested earlier, all four cycles interact to make Earth's function possible. Both atmospheric and hydrologic cycles are influenced, (directly or indirectly and to varying degrees,) by the sun. The interaction is especially easy to see in our daily weather. One is more likely to experience these changes outdoors.

In the early morning, well before the first light, birds begin to stir. Their voices can be heard, as they welcome the new day dawning. Some insects may also get into the act.

As light enters the day, one can often see clouds moving across the sky pushed by the wind. This conveniently displays the atmospheric cycle carrying the clouds, which represent the hydrologic cycle.

A more subtle interaction is also taking place. Water vapor is an invisible gas in the atmosphere. When the volume of water vapor increases, it can eventually saturate the atmosphere. That saturation is represented by the water vapor condensing into water droplets, which begin as white clouds. As the water droplets become denser, the clouds darken. One could almost feel the clouds getting heavier as they become "pregnant with rain", which eventually falls.

The atmosphere facilitates the hydrologic cycle in many other ways. The temperature of the air determines its capacity to hold moisture. We refer the moisture in the air as humidity. The amount of moisture in the air at any given time is a function of what is available. However, warm air has a greater capacity for water vapor than cold air.

Accordingly, if a body of warm air is cooled sufficiently, its moisture capacity may be reduced sufficiently to saturate that body of air. We see this phenomenon at work in the early morning, when overnight moist air has been sufficiently cooled. The resulting condensation produces early morning fog on the windshield of automobiles.

The cycling of water links the hydrologic cycle with the rock cycle most dramatically in rivers, streams, ocean tides and currents. Running water erodes and transports sediments. The transportation of sediment acts like sand paper to smooth and polish rock particles, some of which are carried to the ocean to settle and be compacted into sedimentary rock.

In the rock cycle, rock materials recycle themselves through the three different rock types and the molten form of rock material called magma. Magma is also a product of the tectonic cycle. The twelve tectonic plates, which make up the earth's crust are constantly in motion shifting very slowly.

The plates overlap each other unevenly and jostle each other for position, as they cover Earth. The boundaries between the plates are dynamic. Plates are either converging on each other, or diverging from each other. Where converging plated overlap. The heavier plate subducts beneath the lighter plate and generated considerable friction as their surfaces drag. That friction creates sufficiently intense heat to melt rock material in the vicinity producing magma. This process keeps the tectonic cycle intimately involved with the rock cycle. To give these interactions more clarity in would be appropriate to more closely examine each cycle separately.

CHAPTER 2

The Atmospheric Cycle

THE ENVELOPE OF air that surrounds Earth is what we breathe and cannot live without for more than a few minutes. It is made up of oxygen, (21%), nitrogen, (78%), argon, (0.9%) and carbon dioxide, (0.036%). We and the other animals intake oxygen and exhale the other atmospheric gases. The plants take in nitrogen from other organisms and carbon dioxide from the air. Nitrogen is the building block of proteins.

The air is almost always in motion. Wind is the primary vehicle that distributes heat energy at Earth's surface. When air is heated, it becomes less dense and rises, enabling the surrounding cooler air to "rush in to take its place." The sun heats the land irregularly, causing a tendency for heated air to be unstable. Air pressure is determined by atmospheric density or weight. In other words, less dense air reflects low atmospheric pressure and more dense air creates higher atmospheric pressure. This is why high temperature conditions create relatively low atmospheric pressure.

Because of this phenomenon, wind is generated when a region experiences contrasting temperatures. Wind blows from a high pressure location to one of low pressure.

The sun radiates heat to the earth. In the process, as the sun's rays pass through the atmosphere, only about one-quarter of the rays' heat is initially captured by air. The land beneath it is a much better heat conductor and initially captures about half the rays' heat. Earth's land surface is also better than the air at storing the heat it captures. As such, the atmosphere relies on the stored land heat for its ambient temperature, (the temperature we feel).

Furthermore, as the air closest to the land is heated, it expands and rises, cooling as it moves away from its primary heat source, the land. At the same time, cooler air will become denser and heavier.

Land and sea breezes:

Because land responds to heating and cooling faster than water land and sea breezes are common along coastlines throughout the world. On a warm, sunny day the land temperature will increase faster than the sea. By mid-afternoon, the temperature difference would usually be greatest. Thus the higher land temperature, drawing in a cooler "sea breeze".

At night, especially on a cloudless night, the absence of the clouds causes the temperature to fall rapidly. Because water cools more slowly than land, the latter gets colder than the sea. In the predawn hours, before sunrise, the warmer sea temperature attracts a "land breeze".

Altitude influence:

In mountainous regions, altitude induces temperature changes. The higher the altitude, the lower the temperature is, because temperature falls with increased altitude. At night when the land cools, the colder, heavier air at higher altitudes descends into the valleys.

As the cold air descends into the warmer, more humid air, the cold air condenses in the warmer air to create early morning valley fog. During the day, the sun tends to build up heat in the valley. The hot air rises, cools and repeats the cycle the next morning.

Mountains can take on their own micro climates. It is common for large mountain ranges, (like the Sierras of California), to capture most of the moisture borne by ocean winds on their windward side and render the leeward side a desert.

How does this happen? The warm, moist air from the coast is cooled, as it is lifted into the mountains. It deposits almost all its moisture, (rain and snow), on the windward side, leaving the air cold and dry at the mountain tops. That cold, relatively heavy air, it descends, warming proportionally to become hot and dry on the lower leeward side.

Thus, the California Sierra Nevada Mountains have created the Nevada Desert! Geographers refer to this phenomenon as "the rain shadow effect". The air on the windward side had been cooled at a constant rate of 10 degrees for every 1000 meters ascending. On the leeward side, the air is heated as it descends at the same constant rate of 10 degrees for every 1000 meters. This rate is called the "adiabatic rate" of change, [4].

Another phenomenon of mountainous regions is the automatic change in air density that accompanies altitude change. Where tropical mountains are sufficiently tall, they can have glaciers. For instance, mountains in the tropical Andes of South America have glaciers at the top and tropical forests at their base, as does Mt Kilimanjaro in East Africa, merely one degree south of the Equator

The thin air at high altitudes poses an additional challenge to humans, especially those who do not live at high altitudes. The thinner air makes breathing more laborious. At about 3,700 meters the available oxygen is two-thirds what it would be at sea level. At about 5,500 meters altitude, the oxygen ½ the volume at sea level. This phenomenon presents the greatest risk to mountaineers.

In many parts of the world, isolated areas may have a distinctly different climatic experience from the climate in the surrounding area. This is common in mountainous regions, where the valleys experience many foggy mornings, with temperatures up to ten degrees cooler than the surrounding highlands. Such isolated valleys may have unique micro climates.

Other phenomena can create micro climates. For instance, at the mouth of a limestone cave, a significantly cooler breeze flows outward. The breeze always blows out, because the colder air inside the cave has created a relatively higher atmospheric pressure, forcing air to stream toward the warmth outside.

Mountains are a barrier to the continuation of many travelling weather systems. One dramatic example is demonstrated by the Sierra Nevada Mountains of California, as described above.

Global Climate Zones:

The most dominant high pressure zones are at the poles. High pressure centers can be found in any climate zone and are usually quite transitory. They are also called anti-cyclones, because the winds blow outward from the center, as opposed to cyclones, (low pressure centers), whose winds are drawn inward toward the center.

Nowhere else experiences the blizzard conditions created around the South Pole in the winter. Because this location is in the center of a large land mass, blizzards can often last several months with hardly a break! The polar high pressure zones are much more dynamic in the winter months and feed the mid-latitudes with some of their most extreme cold outbursts.

The hot zone that straddles the equator is a very humid region. Part of the zone is oceanic. The land area it covers is mostly a tropical forest region.

The daily weather pattern of the tropics is hot days and cooler nights, with a normal range of 20 degrees Fahrenheit from nightly lows to daytime highs. There is no distinct summer or winter, per se. The main temperature shifts are experienced between day and night, more so than temperature variations between January and June.

A unique weather cycle is discernable during the hurricane season, (May to October). It is the time of year when the overhead sun is in the Northern hemisphere. Roughly between the sub-tropical high pressure zone and the equatorial low pressure zone a distinct vertical air circulation develops between the subtropical high pressure zone and the equatorial low zone.

The circulation is driven by the air that continues to rise at the low pressure zone. After each thunderstorm, the air continues to rise above the high clouds to become cold and dry. Meanwhile, the air is descending at the sub-tropical high pressure zone. As the high pressure draws down air, it attracts the cold dry air above the low pressure zone, which is heated adiabatically as it descends.

At sea level the descending air encounters the prevailing Trade winds, which blow toward the equatorial low pressure zone. If the air descends in the ocean, the encounter encourages the warm dry air to absorb moisture rapidly, usually causing a tropical depression to develop. According to the analysis of the formation of Hurricane Katrina in August 2005 by The Weather Channel, about half of these depressions develop into hurricanes, most of which fail to make it to dry land.

For high pressure centers to function, they must draw down air from a higher altitude. Thus, the cold dry air above the equatorial region is warmed adiabatically and arrives in the subtropical high region as hot, dry air. Where the subtropical high zone is over land, it tends to produce desert conditions, as in the Sahara Desert of North Africa. Southwest Asia, which includes the Middle East, stretches from the Mediterranean to India. The Dekkan Plateau in central India, the Atacama Desert of South America, the Kalahari Desert of southern Africa, and the desert of Western Australia straddle the Tropic of Capricorn. Like the Sahara to the north, these areas around 30 degrees latitude support Earth's hot deserts.

Other deserts occupy colder regions and are the produce of rain shadow conditions. These include the southwest USA and northern Mexico and the Steppe desert of Central Asia.

In contrast, the oceans adjacent to the tropical desert regions are generally the sources of the tropical depressions at sea that feed the hurricanes of the Atlantic and Indian Oceans and the typhoons of the Pacific Ocean.

For instance, the North America can witness this phenomenon during the summer hurricane season. Their storms begin as tropical depressions off the West coast of Africa. They travel across the ocean toward the Caribbean. If the interior of the North American continent is exceptionally hot, its dominant low pressure center will tend to attract that tropical depression, which may have developed into a hurricane by the time it crossed the Atlantic. The typhoons of East Asia are essentially hurricanes, which were started by depressions off the coast of California and Mexico. They are generally stronger than those experienced in the North Atlantic Ocean and the Caribbean Sea, because they develop over a larger ocean with a considerably greater expanse of warm tropical ocean water to fuel the larger storms.

Climatic Overview:

Tropical, mid-latitude and Polar Regions are determined by physical location. Each is further influenced by the movement of prevailing winds and ocean currents. For instance, on the west sides of the continents, where cold currents exist, wintry weather tends to persist further from the polar regions than comparable latitudes on the east side of the continents, where a warm current exists. In Northern Europe, north of the British Isles, the warm ocean current from the tropics continues north along the coast of Norway to render the Atlantic coast unusually mild winters without freezing weather well into the Arctic Circle.

Continental influences are also significant to climatic conditions. The oceans have such a moderating influence on climates that, where large continents are concerned, the further one is located from the ocean, the more extreme one's climate will be. Thus, places in central Siberia experience the most extreme climates on Earth. Verkhoyansk, Russia, (68 N. Lat 133 E. Long) experiences, on average, a range from about 60 degrees F in summer to negative 53 degrees F in winter. Compare this with coastal Barrows, Alaska, (72 N. Lat 156 W. Long) which has a low of -13 degrees in winter and 39 degrees in summer.

The Eurasian land mass functions as a continent, in that the temperatures in the part of Asia, opposite Barrows, Alaska have similar temperatures. Europe is physically a part of Asia and *not* a separate continent, as some geographic texts have sought to justify. A continent is a large body of land surrounded by water. That condition is the justification for separating the North American continent, from South America. The separation of Europe from Asia is purely political. Europe is on the western part of the Asian continent.

The presence of significant mountain ranges can significantly alter climatic conditions, as we have seen earlier, with the Sierra Nevada range creating a desert in Nevada and with mountain – valley breezes. Large lakes, like the Great Lakes of North America, can produce excessive snowfalls on the eastern side of the lakes in the winter and mild summer temperatures with more rain on the eastern lake side than the interior.

CHAPTER 3

The Hydrologic Cycle

As we worry about the current alleged shortage of energy to meet our global demands, we often overlook the fact that today's most critical global shortage is actually *water*. This shortage is an ever present challenge to people living in arid regions. Any of these regions has been semi-arid throughout recent human history. The water crisis has grown to critical proportions within the last three decades. North Africa, Central Asia and Western Australia have been most critically affected.

To appreciate this dilemma, we must first recognize that fresh water constitutes less than ½ of 1% of Earth's water. Yet, it is consistently available to the human population, where we live. Over 97% of Earth's water is in the ocean. This means that less than 3% of Earth's water at any given time is fresh water, almost 80% of which is currently frozen in the glaciers of Greenland and Antarctica, effectively inaccessible to humans, [5]. As we consume available fresh water, rain and snow precipitate to replenish what we use.

The water cycle may be seen as the continuous recycling of ocean water, because almost all water eventually returns to the ocean. From the ocean, water is evaporated into the atmosphere, condenses into clouds which fall as rain, snow and other forms of precipitation. Much of the precipitation seeps into the ground to help nourish plants and animals alike, as well as supplying ground water. The ground water eventually returns to the surface as river runoff. Vegetation absorbs large quantities through roots, which transport nutrients throughout the plant. The water is eventually expelled through leaves back into the atmosphere by a process called transpiration.

Rain Making: Whenever a body of moist air is cooled beyond saturation, (dew point), water vapor condenses and forms clouds which may produce rain. The air-cooling process is critical to the rain making event. We observed one example of rain making it the discussion of the "rain-shadow effect", (see previous section). The mountain is the "lifting mechanism", (called "orographic lifting"), that provides the air-cooling process, by which the rain is produced.

Besides mountains forcing air to rise and cool, three other "lifting mechanisms" exist. Wherever a cold front converges on warmer air, (common in mid-latitude weather systems), the heavier cold air undercuts the warmer air and force the latter to rise. If the warm air is moist, it may cool to its dew point and produce rain. This is "frontal lifting".

Sometimes, a strong low-pressure system develops. In this scenario, wind is drawn in toward the center from all directions. When the air converges into the center in this way, it is forced to rise in a "convergence" zone. If this is moist air, as it rises and cools in can sometimes produce rain. Much of the summer rain in the mid-latitudes is a result of this process. In tropical regions, as the morning sun heats the air, it rises and cools rapidly, producing thunderstorms in the form of convectional rain.

These four process of rain making: orographic, frontal, convergence and convectional, are caused by physical obstacles to air flow. As one learns to read the clouds, all one need to know about impending rain events is revealed. As a child growing up without weather reports, learning to read the clouds was essential for getting around relatively dry.

Most of the moisture in the air we breathe exists in the form of water vapor. Warm air can retain larger quantities of water vapor than cold air. When the water vapor in the atmosphere is much less than it can hold, the air feels dry. As the moisture increases toward capacity, the air becomes humid. At 100% capacity, the air becomes saturated and the water vapor begins to condense into tiny water droplets.

In other words, the clouds we see in the sky are a sign that water vapor is condensing. The water droplets produced in initial condensation are extremely small. Often it takes thousands of such droplets to combine to form a normal drop of water. As these droplets move around, they combine to form larger droplets. We witness the droplets combining when we observe the clouds darken.

When the condensation process speeds up and cumulus clouds climb upward to form a tall anvil-shaped cloud, a thunderstorm is brewing. Such a formation may give way to a regular or severe thunderstorm. Further developments may produce a tornado, primarily a product of continental mid-latitude weather systems.

People who live in the tropics can expect a pattern of regular afternoon thunderstorms, especially during the rainy season. The convectional rain of monsoon regions produces copious amounts of precipitation from numerous thunderstorms.

If several thunderstorms form close to each other over a warm open ocean, they may organize themselves into a hurricane. Hurricanes require high energy, which comes from the warm water in the ocean. That fuels updrafts in each of the numerous thunderstorms, which fire together like an engine with dozens of very well charged cylinders, to produce the bounty of the greatest rain makers in the atmosphere.

Reading the Clouds:

The clouds can tell us almost everything we need to know about the weather. Altitude and shape are significant in identifying clouds. Because clouds exist at different levels, the direction of cloud movement at different levels can tell us where the winds are coming from, which would indicate the relative temperature of the winds. Where winds of contrasting temperatures come together, especially under humid conditions, it is often a recipe for precipitation.

The low clouds are known as "stratus, stratocumulus and nimbostratus". Middle clouds are "altostratus and altocumulus". The high clouds are "cirrus, cirrostratus and cirrocumulus". At whatever level they occur, their color is probably the most consistent indicator. Darkening clouds indicate the formation of drops of water becoming sufficiently heavy to fall as rain, or snow, sleet or hail.

Stratus clouds are flat and layered. They tend to spread out at the level where they develop. Cumulus clouds are bulbous, (like cotton balls) and tend to climb to form thunderstorm clouds, as they develop. If they do not climb, they generally develop into stratocumulus clouds. Condensation in these clouds is generally slow and can become nim-bostratus clouds, which produce a soft, gentle rain, or drizzle. This type of rain can last for hours, if considerable moisture exists. If, on the other hand, condensation is rapid, the clouds would climb quickly and build into the towering clouds, which we know as cumulonimbus, or thunder-storm clouds.

These clouds climb all the way up to the cirrus level. Cirrus clouds are the wispy, feathery clouds we see high in the sky. They are made up of ice crystals and provide the positive charge that interacts with the negative electric charge in lower clouds made of water droplets to produce lightening. The "nimbo" prefix or suffix in the cloud names refer to rain.

Tornadoes develop from a unique set of circumstances. Three pri-mary ingredients have to be present to spawn a tornado: warm winds meet cold winds, traveling in different directions. The third ingredient is the jet stream travelling directly above this confrontation in the upper atmosphere, [6]. A thunder storm grows out of the mix, often accompa-nied by hail, as the warm rain produced falls through a dense band of cold air, which freezes the falling rain into hailstones. As the cold air en-counters the warm air travelling in a different direction, it causes the air to roll horizontally. The jet stream is so powerful that, if it is sufficiently close, it would tilt the horizontal rolling into a vertical funnel, creating the tornado.

Weather stations can observe these conditions developing and is-sue a tornado warning to residents in the general area. It takes ob-servers in the field, however, to spot an actual tornado and alert the weather station. In areas, where tornadoes are common, the local weather stations recruit and train local volunteers to act as "spotters" during the tornado season.

These spotters are vital to the warning process, since tornadoes are very localized and can develop spontaneously with very little easily observable warning. Weather stations can observe the ingredients coming into place much earlier and usually issue a "tornado watch" to alert residents to look out for signs and call the spotters into service.

The Fog Phenomena:
Fog is a low lying cloud. It is usually formed when warm air and cold air interact in high humidity. Each event displays warmer moist air being cooled under relatively stable conditions. Warm air may be blown into a cooler air mass, or cool air descend into a warm body air, or onto a body of warmer water. This is why we find fog in valleys or over mountain lakes in the early mornings. Sometimes on a cloudless evening, after a hot moist summer day, the rapidly cooling air after sundown can produce a fog.

Recycling Fresh Water:
Even today, less-than-1% of Earth's water is sufficient to supply most of Earth's life, provided that artificial barriers to the access of this precious resource were not created. Because, Earth still rains on the just and unjust, conditioned by natural climatic limits, there is water sufficient for everyone.

The natural recycling of water can take place at many different rates. In most cases, the recycling involves the ocean. The process usually goes from evaporation of ocean water to produce fresh-water vapor, to condensed water in clouds, which are blown to land and fall as rain. Much of the rain drains into rivers, which flow back to the ocean. This cycle may take from a few days to several weeks.

There are many variations on this theme. If it snows, it may pile up in the mountains and be stored for months during the winter, only to be converted when the snow melts in the spring, flow into dams and reservoirs to run off into rivers all summer long. Some mountain glaciers last for decades, as each year's snowfall is compiled and compacted. Thus, the annual layer of fresh snow will make up the top layer. The layer that is melted each year is at the bottom of the compacted glacier.

The glacier melts from the bottom, (as the running water from the snow melt seeps through to the ground and runs beneath the glacier, hastening the glacial-melting process as it runs). This phenomenon causes each annual layer of accumulated snow, (somewhat compacted by its own weight, wind, rain and other factors), to travel downwards. In this way, the current year's snowmelt may contain snow that fell decades before, continuing to be compacted into solid ice by the weight of the many layers of previous snow accumulation.

Artificial Recycling: When industrial water use and extra privileges for the privileged are imposed on the natural process, the water available to the general public becomes severely limited. Mining, industries in need of extensive cooling in their process, sports, such as golf and swimming, plus private swimming pools and green lawns, take a toll on limited public water supplies. The impact is amplified, when practiced in regions with naturally limited water supplies. Desert cities are particularly impractical, [7].

Some Unique Qualities of Water:

Our human perspective on Earth is related to dry land. It is easy to forget that over 72% of Earth's surface is under water, the habitat of more than half of all living organisms, (including the largest).

One unique characteristic of water is its capacity to store the heat it requires to "change state". Water changes state each time it goes from liquid to solid, (ice). When water evaporates it changes to water vapor, a gas. Thus, water exists in these three states: solid, liquid and gas.

The mutual dependency between the atmospheric and hydrologic cycles facilitates the changing states of water and the latent heat that impacts cloud formation. <u>Latent heat</u> is the extra heat required to change ice into water or water into water vapor. In other words, at 0 degrees Celsius, (C), ice melts. The melting process requires additional heat, while staying at the same temperature. We probably recognize the process more easily when water boils. It takes considerably more heat to boil off the water, but the temperature remains constant at 100 degrees Celsius during the changing process. This additional heat is latent and is stored in the water vapor.

When humid air rises and cools to dew point, (the point at which water vapor condenses into water), a unique phenomenon occurs. The water vapor has to release its latent heat as it changes back to liquid water. Remember the rising air was cooling steadily at 10 degrees C per kilometer rise. When the condensation begins, the latent heat is released into the cooling process, slowing down the adiabatic rate of cooling, to about half the rate before condensation[8]. This warming trend is so dynamic in the formation of thunderstorm clouds that the sudden injection of latent heat contributes to the lightning we see.

The Influence of Ocean Currents on Climate:

Ocean currents have a significant influence on the air temperature above the ocean in the general vicinity of the flowing current. A warm or cold ocean current can have a corresponding influence on local weather and climate. San Francisco, for instance, experiences mild summer temperatures, because of the cold ocean current that flows offshore. The prevailing winds called the Westerlies blow inland from the ocean as cooling winds.

For those who live in the San Francisco Bay Area, the influence is dramatic during the summer. Although the water is cold, in the early morning, the adjacent land is colder. The Central Valley of California is hot and its low atmospheric pressure draws in the moist air from the ocean. As this air is brought into contact with the colder land, fog quickly develops and usually remains until almost noon, before finally dissipating. The result is that San Francisco's summer sun is usually available for merely half the day. In contrast, the Central Valley gets full sun daily and their different summer temperatures, (often 15-20 degrees Celsius), reflect this variation. This phenomenon is familiar to those who live along the Pacific coast of North America, (California, Oregon and Washington).

The fact that oxygen is contained in water means that aquatic life is supplied with the life-giving gas. When the water in a lake freezes, it expands between 4 and 0 degree C, contrary to the normal volume-reducing character of cooling up to that point. Furthermore, water freezes from the top down. Thus, as it freezes on lakes, a gap develops between the ice and the unfrozen water. As temperature decline below freezing, the ice contracts, leaving an ever-widening gap of air above the water. This gap makes available a column of air for aquatic animals to breathe when lakes and seas freeze over.

This expansion during freezing causes another common phenomenon. Water that seeps into cracks in rocks and other solid material, causes them to break apart. This action facilitates water absorption into the soil during early thaw.

[This discussion will be continued later in the examination of the rock cycle and in Section III The Stewardship of Earth's Abundance: Chapter on rivers.]

Climate and Vegetation:

The vegetation of each region reflects the climate to which it is subjected. If it is cold for most of the year, the normal growing season for plants and crops would be short, limiting the food crops that would grow there. Even in that region, a wet growing season would probably allow for a more productive growing season than it would in a dry year. Because the tropical region is hot and wet, there is far more plant growth than what we see in Polar Regions.

In turn, the vegetation significantly governs the fauna of that area. Natural forests support a wide variety of both plant and animal life.

Trees need insects to pollinate their flowers and enable the trees to bear fruit, multiply and reproduce. Birds are fed by an abundant insect population and many animals feed off the edible fruit.

The roots of the trees are necessary to hold the topsoil together, so that it is not washed away with every rain. The stabilized soil allows the ground to act as a sponge during rain events. Close to three quarters of the rain that falls on soil covered by trees will be retained in the soil as ground water. The ground water storage will nourish the trees and slowly feed nearby streams, so that these streams may flow year-round.

In this way, each climate generates its own unique ecosystem, based on the plant life native and adapted to that climatic region. The unique ecosystem similarly affects the aquatic and land animals that inhabit the region.

El Nino and La Nina: El Nino and La Nina are complex phenomena resulting from ongoing variations in ocean temperatures in the Equatorial Pacific. The name El Nino refers to the Christ child. Peruvian fishermen used it to describe a warm ocean current that appeared each year around Christmas time. Today we use the term to refer to an unusually strong version of the same event. In the 1600s, it was first noticed as a usual warm current, which changed the types of fish caught at this time. An unusually warm event would occur every ten or so years. Since the 1960s, the frequencies of these events has been increasing steadily, in response to the rising global temperature and repeats itself every few years. Furthermore, the Pacific Ocean is so much larger than the other oceans, that these events are having a global impact on weather[9].

CHAPTER 4

✣ ✣ ✣

The Rock Cycle

THREE BASIC TYPES of rock exist: sedimentary, metamorphic and igneous. All three rock types, when exposed to the atmosphere, running water, and a variety of other disturbances, break down, fracture and break apart, creating fragments we call sediment. Each one of these rock types has been transformed from any of the other types at some time in the history of the planet. Thus, we can recognize a "rock cycle" in the study of the lithosphere, (the earth's crust).

The many processes by which rocks break down are referred to collectively as "weathering". The actual breaking down is more commonly known as erosion and the most common agent of erosion is running water. This is primarily how the hydrologic cycle participates in the rock cycle.

When rain falls much of it runs into rivers, which eventually flow to the sea. Along the way, rivers erode rock material, carry it along and deposit it along its course. Eventually the most finely ground sediment arrives at the sea and is deposited therein.

The river flowing process involves a wide variety of distinct activities, as the river courses its way from its source usually high in the mountains. Where the slope of the valley floors are steep, the river flows swiftly and can sometimes move along boulders, especially during flood stage. In the foothills, the velocity of the stream has decreased and smaller rock particles may be transported by the flow. In a large river, it may flow out of the hills into the higher plains, still at an altitude considerably higher than sea level. Here the slope of the stream bed would be gentler and the sediment being carried would experience alternating periods of transportation and deposition as particles are further broken down.

The usual sequence is as follows: The particles that are too heavy to be transported in suspension by the flowing water get rolled and bounced along the bed of the stream. In this process, rocks constantly collide with each other, cracking and breaking into smaller pieces.

The gravel is made up of the pebbles that continue to massage each other, becoming increasingly smoother in the flow. Their "crumbs" become the fine sand and silt materials that are carried in suspension in the main flow that eventually enter the sea as mud and silt.

The work of the river is time-consuming and changes occur slowly. Thus the landscape has taken a long time to be transformed by a river to what exists today.

The river is very discriminating in what it carries and for how long. First of all it takes running water to move sediment. The greater its volume and velocity, the more sediment it can carry and the larger and heavier the particles it will move. As the flow velocity and/or volume decreases, heavier particles will be deposited. The river bed also loses its steepness along the way and is usually almost flat by the time is gets to the sea.

In periods of abundant rain, rivers may occasionally flood, sending huge amounts of water downstream. In such flood stages, the volume of rushing water may be sufficient to move boulders. In the foothills gravel will flow swiftly in the current, smoothing and polishing shiny pebbles. The pebbles on the beach were probably transported in a flood stage. At the sea the flow stops and all the sediment the river was carrying is deposited.

The ocean is continually in motion, however. Like the flowing river, it continues to break down rock particles still smaller. Some smaller particles become sufficiently light to be picked up in the flow of off-shore currents. Accordingly, rock particles are continually being distributed along the ocean floor, further broken down into still smaller pieces.

The ocean, like the river, continually alternates erosion, transportation, deposition and further breakdown. Because most rivers eventually flow to the ocean, most of the sediment can be found on the ocean bed within a few dozen miles of the land. The ocean proceeds to shift, segregate by size and redistribute this sediment. The sand is deposited on beaches, (being the heaviest of the material in constant motion), allowing the finer, lighter material to be carried back out to sea.

Sedimentary rock is formed when sediment is sufficiently compacted to become "rock-solid". The compaction develops over millions of years, as sediment is compiled where rivers have been flowing into oceans, deposit whatever sediment they were carrying all that time. Over time, the sediment builds up to many miles in depth. The constant movement of the ocean water further redistributes rock particles by size, causing many of the finer particles to precipitate toward the lower layers in the pile. The weight of the upper layers compresses the lower layers until the latter become solid sedimentary rock.

The growing weight of the sedimentary rock and the sediment above it creates pressure, some of which is manifested as heat. This heat energy builds up and accumulates sufficiently to begin to melt the lowest layers of rock. As they melt, the different types mix together to form a new rock type, which re-crystallizes as metamorphic rock.

Igneous rock is formed when magma is cooled and solidified. Magma exists as molten rock material, which originates well below Earth's surface. It generally exists in temperatures in excess of 550 degrees Celsius, (800 degrees Fahrenheit). When magma comes to Earth's surface, its exposure to the atmosphere on land, or the water of the ocean causes it to be essentially transformed. At the earth's surface, molten material is called lava until it solidifies.

Igneous rock crystallizes in many different ways dependent on the type of parent magma from which it solidifies. Magma sometimes exists in huge magma pools beneath Earth's surface. Imagine the Sierra Nevada mountain range in California as a single magma pool that solidified beneath Earth's surface, millions of years ago. The range is over 200 miles long and up to 40 miles wide in places. As the less resistant sedimentary rock above it has been eroded away, over time, the igneous rock, (granite) has become exposed in the form of the majestic, angular peaks visible today.

Each magma pool is unique, being comprised of the particular mix of rock types, further influenced by conditions that existed in the region where the initial melting took place. (This phenomenon will be discussed in more detail in the next chapter.) Different magma compositions produce different types of igneous rock. One major difference is decided by the rate of cooling. Slow cooling magma allows crystals to form slowly. Accordingly, the rock that is formed has larger crystals. Faster cooling magma forms rock with smaller crystals. In some volcanic events, the magma is cooled so fast that crystals do not have time to form. In such situations, a glassy material is formed, called obsidian. Such material is often found at the site of recent volcanoes

As indicated above, magma can cool and solidify within the crust, or mantle of Earth. The rock formed in this way is called intrusive igneous rock. Lava, which is solidified above ground is called extrusive igneous rock. Some extrusive igneous rock is formed in the ocean, usually at mid-oceanic ridges. This rock is generally dark in color and is called basalt. Granite, which is the best known igneous rock, has a lighter color.

Water in Rocks:

We tend to think of rocks as solid and immovable. The last couple pages presented an image of rocks in motion, cycling. The idea of water in rocks seems similarly "unbelievable". But, consider that most of the water that is readily available for human consumption is "groundwater". When it rains, most of the rain water will seep into the soil and drain down to the water table.

To appreciate how groundwater exists, it may be helpful to think of the soil as a sponge. And when a farmer or rural resident bores a well to extract groundwater, the bore goes down into the water table, from which water may be drawn up.

As long as water can travel through the soil and deeper rock strata, that water will seep further down on its own weight. Such rock strata is described as "permeable". Sometimes there exists a layer of underlying rock that is too dense to allow water to pass through it. Such strata is considered "impermeable", and it creates a catchment area below the earth's surface called an aquifer. Some aquifers may cover many square miles and represent a significant reservoir[10].

In such aquifers, the water is such an integral part of the rock structure, that if water is extracted too fast to outpace the water flowing into it, the land surface may collapse in sections. This is particularly true for structures built in limestone regions. Buildings have collapsed into sinkholes created by unregulated or otherwise excessive extraction of groundwater in the area. Parts of Texas and Florida, (and elsewhere,) have experienced such collapses due to local overconsumption of groundwater. Excessive extraction of crude oil from the ground can have similar results, as was experienced in Long Beach, California in the 1930s.

The spongy texture of rock with high groundwater content has been experienced by the parts of New Orleans, Louisiana that now exist below the level of the Mississippi River. That part of the city was built on Delta land, which shrunk when the water was pumped out of the soil to dry it out for construction.

Miners are familiar with the hazard and nuisance of having to constantly rid themselves of the continuous "bleeding" of water from the rocks they excavate. A similar experience has to be accounted for by engineers building tunnels to accommodate subway rail systems. Continuous, round-the-clock pumps are employed to keep the tunnels dry and safe for transporting passengers. Local rainy weather and/or the rock-type in an area may also contribute to excessive ground water.

Limestone, which is a sedimentary rock is notorious for absorbing huge quantities of water. This is one of the most common rock types found worldwide. In limestone areas one can expect to find underground streams, which routinely cause sinkholes to develop at the surface. Building in these areas requires careful examination and restraint[11].

CHAPTER 5

❖ ❖ ❖

The Tectonic Cycle

EARTH'S CRUST CONSISTS of twelve major tectonic plates, which overlap in places to complete cover the planet. Moreover, there is detectable movement on most boundaries. Thus, some boundaries experience convergence, while others are divergent.

Tectonic plates are either oceanic or continental. The continental plates are thicker and the rock combinations are more complex. Although the oceanic plates are thinner, they are more dense and heavier. As a result, when an oceanic plate and a continental plate converge, (as they do at the rims of both the Atlantic Ocean and the Pacific Ocean), the oceanic plate subducts beneath the continental plate.

Tectonic movement is essential for providing magma, from which igneous rock is formed. Magma is the product of rock material melted. The composition of any pool of magma is unique, in that it is made up of the various types of rocks in the vicinity of the heat that melted the rocks. Water in the rocks and the heating process liberate many different gases, which also become an integral part of that magma pool.

Pools of magma exist in various locations within the body of the planet. They are often well in excess of 1,000 degrees C. In these underground hot pools, the excessive heat continually melts the surrounding rocks. Since heat rises, the magma melts the rocks above is added to the pool in ways that cause it to work its way upward toward Earth's surface. This is the source of many of today's volcanoes.

When an oceanic plate converges on a continental plate, the subducting oceanic plate grinds against the continental plate, generating considerable heat from the friction. The friction continues to build up heat as the oceanic plate pushes into Earth's mantle. Eventually that heat melts adjacent rocks to create a growing pool of magma. Most of the volcanoes along the "Ring of Fire" surrounding the Pacific Ocean are formed by this process, including the volcanoes in the Andes of South America and those of the Cascade Range of North America.

Diverging plates, are another common source of new magma. In both the Atlantic and Pacific oceans the oceanic plates in the middle of each ocean are diverging, leaving a prominent mid-oceanic ridge beneath the sea.

These are dramatic active surfaces under water. As gaps begin to appear at the boundaries of the diverging plates magma oozes out to fill the gaps. The result is emerging magma building up from the temporary opening, like toothpaste being squeezed out of a giant tube, which solidifies almost instantly into blobs of lava. Still liquid on the inside, these lumps pile up on each other to look like black sand bags. Thus, they are called "pillow lava".

Such shape identify volcanic activity on uplifted ocean beds. Such stacks of pillow lava are prominent along the Pacific coast of California. The Marin Headlands in the San Francisco Bay Area offers many such examples existing close to the San Andreas Fault. [See later discussion on Mountain Building in the Chapter 9 on Earthquakes]

In the Mid-Atlantic Ridge the greatest activity is in the north, where the buildup of lava has emerged above sea-level to form Iceland. The Hawaiian Islands represent another type of volcanic buildup, creating a chain of islands.

Where adjacent continental plates diverge, the emerging new land is on dry land and lava solidifies in a wide variety of forms. The most common landform is a rift valley, of which The East African Rift Valley is the most dramatic. It stretches from the Red Sea in the north to Mozambique and is a current example of the most rapid divergence on Earth. The landforms range from drowned valleys like the Red Sea to trenches on land, filled with water, like Lake Tanganyika and Lake Nyasa, to mountains like Mt. Kilimanjaro and many other mountains with crater lakes.

CHAPTER 6

Seasons and Landscapes

THE LANDSCAPES ON which we live have been formed by a combination of activities and events initiated by these four cycles. Our local climates have helped to shape the land. The rock cycle has provided the unique rock characters which influence how weathering affects shapes, while tectonic activities move the plates, which alter land masses.

Our choice of a place to live is also influenced by the combination of plants and animals. However, the use of the land and the plant and animal population may have changed. In many parts of the world the indigenous natives used the environment differently. More of the plants and animals were indigenous and their relationship to the natives, was often more complimentary.

We have a symbiotic relationship with all of life on Earth. The insects, (which we tend to see predominantly as pests), pollinate most of the plants that provide us with food, directly, or indirectly. Many animals eat plants we would not eat and extract nutrients which we, in turn, get from them, when we eat those animals. We eat at the top of the food chain and our ancestors have learned much about how and what to eat from trial and error and from what other animals ate around them.

Where we choose to live is governed to a great extent by the weather and/or where we have been forced to live by others, who have some measure of authority over us. Nevertheless, we adapt to the place where we find ourselves.

In fact, we are often far more adaptive that we realize. The majority of us have had to move to very different parts of the country, or other parts of the world. In such cases, we are challenged to adapt to different lifestyles. Often we are forced to do things we never expected to be able to. Many of us travel to other places and are able to find vacation resorts that cater to our lifestyle.

When we venture into the lifestyle of the natives, we are often surprised to find their differences quite appealing, or appalling. By venturing outside our comfort zones, we often discover other ways in which we can broaden our own lifestyle horizons.

Most of us, who live in the USA, are exposed to unusual weather patterns several times during the year. This is a country with a wide variety of landscapes is the only country on Earth where every climate on Earth is represented, (from tropical Hawaii to polar Alaska). Thus, we can travel to any climatic zone without leaving "home".

We also would notice, (from watching weather reports on TV), that the wind can transport the weather from one place to another. And, the wind does not seem to respect national boundaries. Thus, the cold winds that bring unusually cold weather to the Mississippi valley during the winter may come from Canada, rather than Alaska. The Pacific Coast is more likely to be influenced by Alaska's cold weather.

The weather gives us a good example of one of nature's rules. Whatever influences another phenomenon is usually also influenced by it, as well. Thus, even as the cold winds from the north bring unseasonal cold to Georgia and Florida, those winds also warm up as they move south. As such, an air mass that passes through Minnesota with temperatures below zero, may become 30 degrees F. by the time it gets to Georgia, being warmed by the land over which it travels.

Topography can also influence the weather in remarkable ways. During the spring, an occasional wind, called the "Chinook" descends from the Colorado Rocky Mountains into the plains of Nebraska and Kansas, bringing surprising warmth to the land. This is because, the cold winds in the Rocky Mountains descend quickly and are adiabatically warmed to arrive on the plains significantly warmer than the surrounding atmosphere. This is the same phenomenon we discussed earlier that allows the Sierra Nevada Mountains of California to render Nevada a desert.

The distance from the ocean can also influence the climate of a given place. Thus, Kansas and Nebraska have colder winters and hotter summers than places at the same latitude but close to the ocean, because the moderate temperature of the ocean keeps adjacent locations from experiencing extremes.

Learning to Live in Harmony with Nature:

Understanding the four cycles and Earth's lifestyle, as expressed in these cycles would allow us to benefit more from what Earth has to offer. Such an undertaking is by no means easy. Our modern lifestyle is so disdainful of nature that we have grown up systematically ignoring many of the things of nature, which affect us.

The weather is a case in point. When we do consider things like the weather, it is to avoid the aspects of it that inconvenience our lifestyle of ignoring it. When it rains, we call it "bad weather", the only ongoing source of fresh water readily available to us!

Today's greatest, most critical shortage is water. Until the 1960s the planet's climate was virtually unchanged from the turn of year 1 of this era. However, the global population had grown from about 200 million to 3 Billion by 1960. The world saw its first million around 1804. The population had grown to almost 7 billion in 2011. Around that time, the urban population had surpassed the rural population.

The urban modern lifestyle is particularly wasteful of water. Here are some examples:

- Run the water in the sink until it runs warm, with no effort to catch the pre-warm water for other use.
- Run tap while brushing teeth.
- Wash dishes under a continually running tap.
- Leave hose running while washing car.
- Ignore a leaking faucet.
- Leave tap running, while doing several tasks around the kitchen.
- Start running a bath, turn away and forget the bath.

Many major cities throughout the world are critically short of water. In the industrial countries the urban populations generally assume that water is almost "free". For those, who actually pay a regular water usage bill, it is often relatively cheap. Yet, the amount of fresh water available is virtually the same as it was two millennia ago. The resulting water shortage merely accentuates the current climate crisis.

Chapter 19 goes into detail about how we actually use water. A dam in one country could cutoff water in another country[12].

The next section looks closely at our modern lifestyle and its limited capacity to protect itself from most major natural events. That inability is largely the result of our preference for individualism over cooperation in community.

We have chosen to compete with nature, rather than cooperate. Natural events are much more predictable than we think. We are not finding that out, because of our negative attitude toward nature. Accordingly, we ignore too much of the natural activity going on around us daily.

Notes

1. Ambient temperature refers to the temperature in the surroundings of the subject location.

2. For example: If I am located at 10 degrees North Latitude, as the overhead sun travels north toward the Tropic of Cancer, the day it is overhead at 10 N. Lat., will be sometime in April. The days will shorten as it continues to approach the Tropic of Cancer. It arrives at the T. of Cancer on June 21st and promptly begins its return journey to the Equator and the Tropic of Capricorn. From June 21st until it arrives at 10 N. Lat., the days will lengthen. When it arrives at 10 N. Lat., (some time in August), it should register a day as long as that longest day in April. After the appointed day in August, the days will shorten, until the sun is overhead at the Tropic of Capricorn, December 21st, which will be the shortest day of the year for locations at 10 N. Lat.

3. For instance, Seattle has its longest day on June 21st, as does San Francisco. However, because Seattle is ten degrees north of San Francisco, its longest day is approximately three hours longer than San Francisco's. Similarly, on December 21st, Seattle's shortest day is about three hours shorter than San Francisco's.

4. Adiabatic heating and cooling of air temperatures occur at a steady rate, when air is heated, or cools. See Edward J. Tarbuck and Frederick K. Lutgens: Earth Science, 11th edition, c2006, Pearson Prentice Hall, pp. 473-474.

5. See the distribution of Earth's water, ibid., Christopherson, 7th edition, p. 180. Consider that much of Iraq is semi-arid.

6. The Jet Stream is a high velocity current of air, which flows between a cold air mass and an adjacent warm air mass in the upper atmosphere. It travels in roughly the same direction as the prevailing westerly winds of the mid-latitudes. Because the air masses are almost always in motion, they cause the Jet Stream to meander between the contrasting air masses. Because of the high altitude, the absence of resistance encountered by wind at ground-level, the Jet Stream builds up enormous speed. Air planes try to ride the current on their eastward flights and avoid it on their westward flights.

7. Cities, by their nature use large quantities of water. Seven of the ten cities in the USA expecting to run out of water soon are in semi-desert to desert regions. See Stockdale, Charles, et.al.. "The Ten Biggest American Metropolitan Areas Running Out Of Water," by Charles B. Stockdale, Michael B. Sauter, Douglas A. McIntyre, November 1, 2010.

8. The dry adiabatic rate of cooling is 10 degrees Celsius for every 1000 meters of elevation. When condensation begins to occur, latent heat is injected into the process, reducing the adiabatic rate

to an average of 5 degrees Celsius per 1,000 meters. Ibid., Tarbuck and Lutgens, pp. 473-474

9. El Niño, warmer than average waters in the Eastern equatorial Pacific (shown in orange on the map), affects weather around the world. El Niño and La Niña are opposite phases of what is known as the *El Niño-Southern Oscillation* (**ENSO**) cycle. The ENSO cycle is a scientific term that describes the fluctuations in temperature between the ocean and atmosphere in the east-central Equatorial Pacific (approximately between the International Date Line and 120 degrees West). La Niña is sometimes referred to as the *cold phase* of ENSO and El Niño as the *warm phase* of ENSO. These deviations from normal surface temperatures can have large-scale impacts not only on ocean processes, but also on global weather and climate. El Niño and La Niña episodes typically last nine to 12 months, but some prolonged events may last for years. While their frequency can be quite irregular, El Niño and La Niña events occur on average every two to seven years. Typically, El Niño occurs more frequently than La Niña.

10. One aquifer, (High Plains Aquifer), stretches from South Dakota to Texas and includes parts of Wyoming, Colorado and New Mexico. p. 263, Christopherson, 5th edition.

11. Construction on a limestone landscape is vulnerable to a potential underground collapse, creating a sinkhole at the surface. In 2004, Lake Chesterfied in one of the suburbs of St. Louis suddenly drained in three days, leaving a gaping sinkhole about 20 meters across at one end of the lake, (ibid., Tarbuck and Lutgens, pp. 147-148. See also, discussion on Wolf Creek Dam in Kentucky in Section IV

12. For decades, the dams on the Colorado took so much water from the river that it would run dry before it got to the ocean in Mexico. The following two articles describe the process, whereby

a new agreement was established, which now allows water to flow through the lower course of the river to reach the ocean. Read: The Colorado River Runs Dry, by Sarah Zielinski, Smithsonian Magazine, October 2010. And Colorado River begins flooding Mexican delta, Associated Press, March 27, 2014.

I I

Natural Events

Dramatic Displays of the Four Cycles in Action

Earth speaks to us in many ways.
Earth's voice is ever present,
even in natural events
. . . . . if we would only *LISTEN!*

NATURAL DISASTERS ARE among the natural events most feared by people throughout the world. Each natural event is a pivotal part of one of the Four Cycles. Such events today would not be disasters, (for the most part), had our current land use practices been developed in harmony with Earth's lifestyle. Furthermore, it would help us individually and collectively, if we were actually living in harmony with Earth. Stewardship teaches us the harmony of Earth's lifestyle.

What is Earth Stewardship?

Earth stewardship is so basic that all animals do it! Each animal is taught by its parent how to survive in its natural environment. First the new-born learns its natural form of mobility. For birds, it means learning how to fly. The new-born is initially fed by its mother, sometimes with help from the father.

Once its mobility has been mastered, the next learning is to feed self. For some species, hunting skills are essential. Some hunt other animals, while others hunt and gather plants. Some predators rob weaker competing predators of what they caught. Still others scavenge the leftovers of other hunters, having eaten their fill.

Then there are the plants. They are the most abundant of living organisms. They purify our atmosphere. They capture proteins from sunlight to pass on to us. They stabilize the soil to enable us to grow crops in it. They are the source of every medicine we need for healthy living. They feed the entire animal kingdom. Some plants protect other plants in their infancy to enable them to mature healthfully. Some plants provide support for healthy vine growth.

The birds sing in each new day. They signal to the animal kingdom the coming of the rain. Many birds are active pollinators. They cull the insect populations.

The insects are our primary pollinators. They reproduce in abundance and feed the birds. In fact, numerous animals and plants reproduce in abundance and, in so doing, feed others in the animal kingdom. Ultimately, the expired carcasses of all plants and animals return to the soil to nourish it.

Herein lies the cycle of perpetual sustainability. All life on Earth plays its part in the sustainability of the life of the planet. In a real sense, stewardship is the function of taking care of one's environment. For Earth's organisms, living in harmony with their surroundings constitutes taking care of it. Their primary response to life is a direct response to their environment. They give to it and take from it in spontaneous response to their respective ways of life. In traditional societies, people live off the land and express a primary response to their environment, as an integral part of their respective life expression.

It cannot be overemphasized the extent to which a positive, endearing response to the environment fuels harmony. After a heavy rain, the birds are the first to lift up their voices in excitement over what the rain has brought for them. Their joy is palpable!

This has been my experience living in tropical forest habitats. Tropical forests are home to the densest animal populations on Earth. The place is alive with the voices of nature. The atmosphere is alive with the fragrance of flowers and fruit. Competition exists. However, the prevailing productivity of each other's efforts in a natural abundance, which allows for the feeding of all organisms in a show of genuine interdependence.

Earth Emits a Constant Array of Messages:

Earth issues a continuous array of messages in a variety of animal languages. That is why in traditional societies people put so much effort into living in harmony with the local animals. Human senses are under-developed in comparison with many organisms in the animal kingdom. Our sense of smell, hearing, seeing, tasting are all inferior. Our vocal capacity is probably more flexible than most in the animal kingdom. However, their mental telepathic scope is far superior to ours.

Modern society has denied and suppressed its spiritual capacity, especially since the dawn of the colonial era. Its heightened appeal for independence is a false priority. Its preoccupation with elevating a will to compete for "the best" has taken us away from the more natural and interdependent appeal to stewardship, with its incidental product of sustainability.

Wherever native populations have been displaced by European cultural suppression, traditional spiritual customs have been discouraged, if not outlawed and repressed. All organisms struggle for their own survival. However, their respective survival efforts complement each other and contribute to an inescapable web of cooperative interdependence, which is hard to ignore, when the primary focus is survival and the protection of self, family and community.

In traditional society, there has always been an important instinct for communal survival. In fact, the village, with its emphasis on the community and the common good, has always directed governance toward the service of community.

The attitude of community and cooperation is better displayed by domestic pets than by modern society. The colonial fixation on competition in an arena of protected inequality has driven modern society into a form of governance, which seeks the common good through a mentality of "survival of the fittest". The instinct for cooperation is discouraged.

Yet, a growing body of research on domestic pets has suggested that they can often read the thoughts of their masters and generally respond with vivid expressions of cooperation. Cats and dogs are particularly adept. Research in pet dogs reveals an amazing capacity to anticipate the thoughts of their owners[1].

As such, traditional societies have been able to derive adequate warnings of impending natural events for thousands of years prior to the dominance of our modern civilization. Because our distant ancestors planned most of their ongoing activities around their active observations of weather changes, the patterns of flow in their local rivers, the seasons, the behavior of local animals and many other natural phenomena. They were always more and better prepared for respective environmental changes than we are today. Thus, they could use the superior sensitivity of these animals to benefit from the animals' anticipation of natural events.

<u>Why do These Events Become Disasters</u>? In modern society, we actively ignore our physical environment. Our instinct for acquiring "the competitive edge", often at all cost, reduces the opportunity to seek cooperation, even when it makes most sense. Moreover, modern society is quite hierarchal, with a full assumption that we have nothing to learn from "the lowly animals".

Most natural events follow quite predictable patterns. Traditional societies observed these patterns closely and based most of their ongoing activities on such observations as weather changes, the seasons and many other related natural phenomena. Their genuine humility left them better prepared, (than we are today), for the natural events of their respective regions.

In addition, the people in traditional societies were well tuned into the behavior of their local animal neighbors. Thus, they could take cues from the behavior of animals', (and plants), to anticipate what event was probably imminent. So, they could take necessary evasive action accordingly. For instance, the sudden absence of birds warns of an approaching storm that may not otherwise be obvious to a casual observer[2].

Other observations of the surroundings were critical. They developed an intimate knowledge of their immediate physical environment. The animals and plants in their neighborhoods were familiar to all. Where they chose to live reflected such knowledge and awareness.

Our modern lifestyle is not based on acute personal, or communal awareness of the natural environment. Almost everyone talks about the weather. However, we seem oblivious to the unique inter-relatedness of the weather events and our daily needs.

Today people build their homes in flood plains and are surprised when those houses are threatened by floods. This is because decisions to build homes and locate urban development are driven by marketing forces, with little regard for environmental limits and related risks to public safety. In fact, more than a half of the global population live in urban areas. Yet, few live with a sense of preparedness for the natural events of their region, which carry significant risk of harmful consequences.

The following several chapters will describe common natural events, outline related warning signs. Their discussions would explain the customs of traditional societies, which allowed them to avoid the disastrous consequences of such events to their respective populations. A common theme creating most disasters is that the European colonial psyche tends to see nature as a problem to defeat, or an obstacle to overcome. Such a mentality is resistant to the idea of cooperating with nature. Indeed, much of modern technology is designed to "subdue nature" in keeping with a popular misinterpreted Christian principle. Ironically, it comes from a mistranslation of the concept of Earth stewardship put forth in Christian scriptures[3].

CHAPTER 7

❖ ❖ ❖

Magma and Volcanic Activity

THE "NATURAL DISASTER" most closely related to the tectonic cycle is the eruption of volcanoes. Their impacts can be devastating to large numbers of people when erupting in densely populated regions. Eruptions around the world are quite uncommon, when compared with other major events, such as hurricanes, tornadoes, earthquakes, major floods, forest fires, etc. Volcanic activity is more complex and host far more potential hazards than any other natural event, as we will see below.

Some eruptions are violent and explosive, like Mt. St. Helens in Oregon in 1980, or Krakatoa in Indonesia during the 1880s. Others are quite gentle, like the volcanoes of the Hawaiian Islands. The scale of any of these events can vary greatly. For instance, Mt St. Helen's sent ash to the harbor of New York. Mt Pinatubo in the Philippines was over five times as strong and colored sunsets globally for weeks after its initial eruption. The Krakatoa eruption was so violent that its explosion could be heard half-way around the world!

Early European history recalls several major volcanoes in the Mediterranean, such as Etna and Vesuvius. Native American legend tells of a volcanic eruption, which has been confirmed by geologic record to have occurred around 7,700 years ago in the Klamath basin of eastern Oregon to create Crater Lake.

Nevertheless, the biggest volcanoes in geological history are many times more devastating than anything experienced in recent human history. The volcanoes that grab our attention and fear are the explosive ones and the ones that spew gasses, which surprise and kill people. Others, like the Iceland eruption of April-May 2010 filled the lower atmosphere with clouds of volcanic ash, and interrupted air traffic throughout much of Western Europe for two weeks. A similar eruption in the early 1960s in Costa Rica lasted over a year and spread thick blankets of ash throughout the basin of several hundred square miles, in which the capital city, San Jose is located.

The volcano, Soufriere, which erupted in St. Vincent in the Eastern Caribbean in 1902, sent a pyroclastic flow of poisonous gases down the slope into the town below. Before anyone realized what was taking place, they were all suffocated. "There was no warning!" said the colonial officials, who had little knowledge of the natural warnings that are well known by the natives, who lived off the land.

The most common warnings are small tremors on the flanks of a volcanic mountain. The tremors are caused by the swelling of the magma chamber, which is forced to expand in response to the expanding gases in the hot magma, as well as the expanding magma, itself. The tremors increase in frequency as the eruption draws near.

An eruption may release decades, even centuries of pressure built up in the magma chamber. The ancestors lived on the flanks of volcanic mountains to benefit from the exceptionally fertile volcanic soil that grew their crops in abundance. They appreciated the cycle that renewed volcanic soil after several generations.

The story of Crater Lake in Oregon was handed down by the natives of the Klamath basin, who called themselves the Makalak, (people of the marsh). They saw the physical and spiritual world as one. Each animal was seen as an individual reflection of creation, with which they shared Earth. All forces of weather and geological change were, for them, the embodiment of powerful spirits[4].

For the Makalak, Mt. Mazama was the center of their world. They believed that two competing spirits were fighting for control of their land. Skel was the "above-world spirit", who lived in the marsh country, represented goodness and beauty. His messengers, (such as eagle, deer, beaver,) were the most beautiful creatures of the marsh, when they appeared in visible form. Skel was the spirit of light.

The spirit of darkness was Lau. He lived in the mountain, Mt.Mazama, which stood about 14,000ft high. Lau was angry and wanted to come out and destroy the people. Skel held Lau imprisoned in the mountain. For years Lau growled and shook the mountain. When Lau become very angry it shook the mountain violently. The people fled in terror, as the battle raged in full force between Skel and Lau.

Lau blew off the mountain top and tried to come out. But Skel was stronger. He pushed Lau back into the mountain and collapsed the top of the mountain back into the crater, covered it and sealed the entrance. For many years afterwards Skel allowed it to snow and rain until it filled the crater with beautiful blue water, so that Lau could no longer enter the world of the people of the marsh. They called it "The Lake of Blue Water". It became their most sacred place.

It is noteworthy to remember that the Makalak story was handed down by their oral tradition. They dated the Mazama eruption to be over 7,000 years ago. Our modern geological estimate placed the volcano to be around 7,700 years ago. The story also indicates to us that their lifestyle was earth-focused. When the mountain grew unusually "angry" they took it as a warning that it was time to evacuate their regular settlement. We use our scientific learning today to deduce that the increasing frequency of local tremors on a volcanic mountain indicates an approaching eruption.

The Humility of our Science and Technology:

We feel we need to know the exact day and time of day of the expected eruption. That is still an unrealistic expectation, even for our science. We never need to know exactly when. At a certain point, one needs to assess for oneself: I will have to evacuate, or die. I need time to evacuate. I want to live, so I go now. Basic, human-scale fear will tell me when it is time.

We cannot go back to living like our ancestors. Too much has changed to recapture the things that worked for them. It will not match.

However, as we continue to observe what went wrong with our approach to life and explore potential alternatives, we will find solutions. In fact, we already have found solutions. Our prevailing global mindset prevents us from making the critical decisions necessary to promote decisive action, where we know it needs to take place. This particular dilemma will become more apparent later in this study, (see Chapter 21 on the Crumbling Infrastructure).

Expecting the Eruption

The popular concept of predicting natural events is unrealistic. Earth is not a machine run by our clock-calendar time system. Each of its events is the product of a host of variable forces, each of which flows in its own timeframe. The sequence of inputs that produce a volcano is unique for each eruption. Volcanologists have mastered sufficient fore-knowledge to predict within a few days that an eruption is imminent. This was the case when waiting for the Mt. St. Helens eruption in Washington in 1980.

From the time of the initial volcanic activity is detected to the actual eruption, weeks or months may pass. For the safety of the people in harm's way, early evacuation is necessary and patience is required. This was the case in the village of Armero in Colombia in November 1985. The mountain, Nevado del Ruiz stands some 70 Km from the village, which was built on the lahar that completely covered an earlier village occupying that space in 1845.

The mountain had been showing signs of volcanic activity beginning in November 1984, and was of significant concern to national scientists. A young scientist, Marta Calveche, who was studying the path of previous lahars became alarmed to learn from her studies that the village of Armero was built on such a lahar, which had come from the mountain's previous volcanic eruption of 1845. A few hundred villagers had perished at that time. The new settlement, however, had grown into a town of over 20,000 since then and was even more vulnerable.

They contacted local town officials to encourage evacuation. But they had heard of the concerns of the scientists for months and nothing had happened yet. The town officials were not convinced. They said they would need a guarantee of a specific date in order to announce an evacuation.

A few weeks after the time that the scientists had recommended the evacuation of Almero, the mountain, Nevado del Ruiz erupted in the night. It took a few hours for the resulting lahar to travel the 70 kilometers along the valley, to awaken the settlement with mud flowing through their houses. People at the edge, where the mud was not too deep, could wade through it to safety. However, many were not so lucky. Some were in houses and could not get out. Many found the mud too deep to wade through and died in the open, in full view of helpless rescuers. The drama continued for weeks as television crews carried it worldwide. By final count, the disaster took over 23,000 lives[5].

The Mt. Pinatubo experience in the Philippines six years later learned from the Armero disaster. The progress of its buildup was well monitored. People were evacuated within a radius of 30 Kilometers and very few people were killed by the initial eruption. It was a much larger volcano and resulted in less than 800 deaths, because the people were ready to evacuate. They started almost ten weeks before the eventual eruption. Smaller eruptions began about two weeks before and a wave of 25,000 to 30,000 evacuated. By the time the mountain had its full eruption on June 15 1991, a total of 58,000 had evacuated.

The eruption occurred, exploding into a monsoon thunderstorm. Such thunderstorms generate huge amounts of rain. It is not uncommon to have many feet of rainfall in a day. Of course, the excessive heat from the volcano only adds to the rapid updrafts of heat that enable thunderstorms. The downpour ignited an enormous lahar that rushed down the many valleys that drain the mountain.

But, it wasn't the lahar that killed the 800 people mentioned above. Most of the deaths were the result of roofs collapsing under the weight of the huge volume of volcanic ash deposited far from the vulnerable valleys. This was the second largest and most violent volcano on Earth since Krakatoa in the 1880s. Of course, the scientists in the Philippines could not predict the day the mountain would erupt.

Neither could the scientists in Colombia. The community certainly had the technology to identify the warnings and avert such a disaster. However, our skewed sense of time and its related priority structure will continue to cause the general public to be sacrificed for the convenience of corporations, who tend to consider the welfare of communities far less important than their economic wellbeing.

Non-Eruption Hazards

Even without an actual eruption, gas leaks are common around the craters of composite volcanoes. This poses many different potential problems. Moreover, volcanic mountains can develop a network of internal channels with numerous vents, from which poisonous gasses, (such as carbon dioxide, which is odorless), may leak. Carbon dioxide is heavier than air and tends to settle close to the ground, when it leaks. If the vent is in a crater lake, the gas may build up to dangerous proportions, without being detected, and would cling to the surface of the water.

If it is escaping into a valley or depression it would seek the lowest area to accumulate. On the flanks of a mountain, the risk is particularly high. With magma building up in an active composite volcano many fissures may develop on the mountainside, spewing out gasses, such as nitrates, sulphates, chlorides, fluorides, carbon dioxide, etc. Because small children closer to the ground, they are most vulnerable in any hollow area.

For instance, on the flanks of Mt. Nyiragongo, near Goma in the Congo, which erupted in 2002, children died while playing in a hollow near a school. A fissure had developed in that depression at the edge of the school yard and carbon dioxide was leaking[6].

In 1986, an entire village was wiped out by poisonous gases that spilled over the rim of the crater. It happened in the Cameroons, where

Lake Nyos a crater lake on top of a mountain experienced a land slide into lake. Carbon dioxide had escaped from the vent at the bottom of the lake and was dissolved in the water. The landslide had agitated the dissolved carbon dioxide, which released a large quantity, filling the crater. After filling the lake, it spilled over the summit and crept down the mountainside, suffocating all life with which it had come in contact[7].

In effect, volcanoes court a series of potential dangers. Each volcano has its own combination of challenges, some of which do not warrant the risk of continued settlement. As such, in some cases, settlement on the flanks of active composite volcanoes should probably be evacuated completely, when the leakage of volcanic gases is detected. Such prompt evacuations are much easier among people who live in tents or other temporary structures.

The skill of stewardship is particularly critical in the vicinity of composite volcanoes. It should be one of the critical skills taught in elementary school and enhanced in high school. In other words, people, who live among active composite volcanic need to be taught a level of stewardship awareness of their larger surroundings.

A volcanic mountain with a glacier on its top presents still other challenges not necessarily dependent on an eruption. The heat from an internal magma chamber beneath the ice cap may gravitates close to the surface and melt large sections of the ice and send torrents of water down the mountainside, creating a lahar. It then would flow like watery cement and hardens to a texture of concrete, when it stops flowing. Many experts fear such an event may happen on the flanks of Mt. Rainier in Washington, USA, at any time, with little additional warning[8].

The cities, towns and other settlements are located on the deposits of previous lahars should be so named to broadcast appropriate warning. Many such settlements exist on the flanks of Mt. Rainier. The mountain contains an active magma chamber within, while supporting the largest mountain glacier in the Cascades.

Parts of the glacier have been weakened by magma chambers expressing excessive heat in recent years. It is possible that such a heat build-up could super-heat a sufficiently large section of the gla-cier to trigger a spontaneous lahar created by a sudden glacier melt. Monitoring this glacier becomes a critical ongoing stewardship task.

On tropical mountains, in a monsoon climate, heavy rains can melt the snow, creating excessive runoff, which would produce voluminous lahars. Mt. Pinatubo, (in the Philippines), had no snow on its flanks, but when it erupted in 1991 into a monsoon downpour. The resulting lahar affected several villages on the rivers downstream from its flanks. It erupted again in 1992. Although the eruptions of 1991 and 1992 killed hundreds, most of the actual devastation has come from the series of lahars that followed for several years after the initial eruptions. Over that period, 19 villages had to be abandoned as over 8,000 homes were destroyed and close to 75,000 other were damaged. However, the people heeded the warnings and evacuated many times during the monsoons, for the next several years.

Factors Influencing the Character of Magma:

A magma sample is made up of the rock materials melted to form it. Any water or other gasses that may have been present at the time of melting the rock sample in question have been incorporated into the molten mass. This gives each magma sample a unique character and composition.

Magma samples are so unique that volcanologists can tell from a stray deposit of volcanic ash, which volcano it came from. This factor has been vital in tracing how far volcanic ash has been dispersed from a given volcano. Volcanic ash in ice core samples from Greenland and Antarctica has been used to date obscure prehistoric volcanic eruptions.

Magma is molten rock material found in pools beneath the earth's surface. When this molten material emerges at the surface, it is called lava which creates "extrusive" igneous rock, when it solidifies. Magma that is solidified beneath the surface is called "intrusive" igneous rock.

Extrusive igneous rock on land is generally lighter in color, because its exposure to the atmosphere causes it to be oxidized on its surface.

The lava that solidifies in the ocean is generally basaltic in nature and has a darker color than lava on land. Granite is the most common igneous rock found in North America.

Volcanoes are fed by large pools of magma beneath Earth's surface. Magma that comes from composite volcanoes, like Mt. St. Helens, Mt Fuji, Krakatoa tend to be very explosive, because they have a high silica content. The volcanoes in Hawaii are low in silica. They are basaltic and their lava cools very slowly.

Mountains that are built from volcanic eruptions usually develop characteristic shapes. Those built from lava that cooled rapidly tend to be steep sided. Those whose lava cooled slowly tend to have gentle slopes. Other variations exist.

Furthermore, the igneous rocks that are formed by fast-cooling material create rocks with a fine texture, because the crystallizing process did not allow the crystals time for much growth. Slower cooling causes crystals to grow large, producing rocks with a more coarse texture. When cooling occurs so rapidly that crystals do not have time to form, the rock is glassy, like obsidian or volcanic glass.

Another characteristic of explosive volcanoes is magma with a high content of gasses trapped in the liquid. The existence of gasses often causes pressure to build in the magma chamber. Where the volcano has had previous eruptions, the plug that solidified in the main vent is usually the weakest point in the sealed chamber.

Pressure in such a chamber can build up for many months, or years, before the volcano eventually erupts. However, if the seal is broken before sufficient pressure has been built up to explosive levels, gas and steam may escape sufficiently to relieve the pressure.

Volcanic Structures by Shape:
Under-sea volcanoes tend to have a low silica content and low viscosity. These volcanoes are seldom violent. The lava that comes from these vents solidifies quickly in the surrounding water. They form as blobs and

pile up on each other to appear like sacks of cement. They are called "pillow lava" or "pillow basalt" and can often be seen above ground near tectonic plate boundaries.

Many of the volcanoes that erupt on land have a high silica content. Because of this composition, their eruptions are violent and explosive. Lava that flows from them is generally more viscous and the mountains they form are steep-sided because the lava cools and solidifies quickly and does not flow far from the vent before becoming solid. A volcano that has been formed by a single eruption is known as a **cinder cone**. When steep-sided volcanoes are made up of alternating layers of lava and volcanic ash, they create taller mountains called **strata volcanoes**.

Slow-cooling basaltic lava builds gentle-sloping mountains, as we see in the Hawaiian Islands. This type of lava comes from magma that is low in silica and is referred to as a shield volcano.

Because some lava flows slowly and has very low viscosity, some flows can spread out for long distances over flat land and be solidified without becoming a hill, or mountain. Such formations are known simply as lava flows.

The most popular volcanoes are the violent strata volcanoes. When one such volcano has accumulated a large magma pool beneath its summit, in which considerable pressure has built up, an explosion may be sufficiently strong to empty the upper magma chamber. In such a condition the emptied cavity may collapse, leaving an extra-large depression known as a caldera. Many craters become filled with rain water, forming a crater lake. However, it does not always form a lake. Some lava flows spread out over the land to cover several hundred square miles. One such lava flow has spread over Eastern Washington and Eastern Oregon, into Idaho[9].

Mega-Volcanoes:

Geological records and complementary research have pointed to evidence that Earth has experienced the enormous power of past eruptions, which exceeds anything experienced in recent human history. Much of this evidence can be linked to spectacular calderas throughout the planet. We mentioned Crater Lake earlier. Over 5,000 feet of the mountain top of Mt Mazama collapsed into its own crater over 7,000 years ago to form Crater Lake. It had been thought, for many decades, to be the largest crater-lake on Earth, until a huge lake in Indonesia was recently discovered to be an actual crater lake.

Lake Toba is now believed to have been formed by a mega-volcano about 75,000 years ago. It is estimated to represent the largest explosive eruption anywhere on Earth in the last 25 million years. Such an eruption would have put so much volcanic ash into the atmosphere that it would have spread around the planet in thick layers. The blanket ash clouds containing sulfuric acid plumes is believed to have remained in suspension in the atmosphere for years, blocking out the direct sunshine. As the clouds circulated the earth, it cooled the planet to a volcanic winter with a worldwide decline in temperatures between 3-5 degrees C, and up to 15 degrees C in higher latitudes.

After such an eruption large amounts of volcanic ash fall to the ground, a small fraction of what makes up the cloud that is blown so high that it remains in orbit around the planet. If that eruption lasts a few days sufficient ash would be in orbit to block out the sunlight for an extended period. 75,000 years later geologists are still finding samples of ash from the Toba volcano.

As stated earlier, each volcano emits ash of a unique composition. This characteristic provided a common ground, which has allowed several scientists to collaborate ash findings to identify and locate Lake Toba as its source. The collaboration was inter-disciplinary. In fact, it was the initiative of researchers, who were willing to share information with people outside their field, who were most effective in collecting valuable cross-disciplinary data[10].

In the Spring of 1994, a geologist from Eastern Illinois University, Craig Chesner, while studying the lake floor of Lake Toba was surprised to find that, unlike most lakes, its bank presented a sheer drop off in parts. Then, on the island in the lake, he found it to be made up mostly of volcanic ash and pumice. So he sent a sample to Westgate. Thus, after several years of research he had finally found his match at an appropriate site. Lake Toba located in far-away Indonesia proved to be the megavolcano, which had deposited the mystery volcanic ash in Greg Zielinski's Greenland ice core sample.

Evidence of other eruptions of megavolcanoes can be found elsewhere in Southeast Asia, Japan, Italy, North America and New Zealand. Four of these sites are still active: Long Valley. California, Yellowstone, Wyoming, Lake Toba and Lake Taupo in the North Island. Each volcano sits on a huge magma chamber, estimated to contain over 200 cubic miles of magma.

To give an idea of its magnitude, consider that although the eruption took place in Indonesia, it deposited an ash layer an average thickness of 15 cm (5.9 in) over the entire landmass of South Asia; at one site in central India, the Toba ash layer today is up to 6 m (20 ft) thick, and parts of Malaysia were covered with 9 m (30 ft) of ashfall[11].

In addition it has been calculated that 10,000 million metric tons of sulfuric acid would have been ejected into the atmosphere by the event, causing acid rain fallout[12]. In short, we have to adopt a new level of environmental awareness and stewardship specifically related to the scope of volcanic activity when we live in the vicinity of volcanoes.

Earthquakes

"Earthquakes do not kill people; buildings do!"[13]

AN EARTHQUAKE IS essentially a sudden movement of large blocks of land along a fault line. Fault lines are cracks that develop from folds which often form when sedimentary rock is exposed to the slow, steady, excessive compression forces of converging tectonic plates. For instance, where an oceanic plate converges on a continental plate, the flat layers of sedimentary rock deposited on the continental shelf begin to fold. Other compression forces on sedimentary rock layers can also cause folding.

The continental shelf is the gently sloping edge of the continental plate that extends under water from the coastline out to sea. It is made up primarily of sedimentary rock brought down to the ocean by rivers and further distributed by the ocean's tidal action. Earthquakes occur only in sedimentary rock beds, usually where complex folding has taken place.

The folding is very much like what happens when someone pushes against the loose end of a rug, causing it to "bunch up". Examples of that folding process are visible in the offshore islands of the Channel Islands off Santa Barbara, Ventura and Los Angeles, on the Pacific Coast of southern California. The coastal range of the California Pacific Coast is also a part of that folding process. How do we know? Sea shells and pillow lava deposits in the hills bear witness to the folding process that lifted the sea floor out of the water[14].

This development is commonly seen along the Pacific Rim and much of the Atlantic coast in the form of coastal mountain ranges and off shore islands. Road cuts in these areas often reveal the characteristic folding of the layers of sedimentary rock, which were initially laid down flat and compacted on the ocean bed of the continental shelf millions of years ago.

Road cuts may also reveal breaks in the folds. When the stress on the folding rock layers builds up to exceed the bending capacity of the rocks a break occurs. Such breaks often become the faults along which further movements produce more earthquakes.

The rock layer under stress builds up energy which is released when an earthquake occurs. As stress builds up, the rock base may experience a break at its weakest point. The break becomes a fault, a line of weakness, along which future earthquake movement will result, releasing stress. Slippage along the fault line may result in either vertical or horizontal displacement, depending on the direction of the force and the primary angle of the fault line.

Most earthquakes occur on one plate or the other, near major plate boundaries. At present, most of the active earthquake and volcano zones on Earth are located close to the plate boundaries that circle the Pacific Ocean, (called the "Pacific Rim"). About 30,000 earthquakes are felt every year, worldwide, of which only about 75 are significant.

Shock Waves:

The energy released by an earthquake travels through the earth's crust in the form of shock waves[15], which radiate from the point of origin of the energy release, called the "focus". Since the origin is always underground, it is identified by the "epicenter", which is located at the surface, directly above the focus. Thus, what we actually feel of the earthquake is the passage of shock waves through the ground on which we stand.

There are essentially two types of shock waves: "surface waves", which we feel, as they travel along the surface of the ground; and "body waves", which travel through the earth's crust. The body waves are the initial waves that radiate in all directions from the focus. When the body waves strike the surface, they begin to travel along the ground at surface level as surface waves, outward from the epicenter.

The body waves are most important to us in measuring and locating the earthquake. Two sets of body waves emanate from the focus. They are the primary, or ("P") waves and the secondary, or ("S") waves. Their importance lies in the fact that they are generated simultaneously at the focus. However, the P-waves travel faster than the S-waves. Their relative speed is constant, so that, when they arrive at a seismic station, the distance from the focus can be calculated. It takes locations from a minimum of three stations to calculate the exact location of the epicenter.

Earthquakes are relatively harmless outdoors in an open field. The problems we have with this natural event all relate to buildings and the mad-made structures we occupy. Unfortunately, we have developed an indoor lifestyle that leaves us vulnerable the damage and occasional collapse of the buildings we occupy. However, technology has given us considerable skill and expertise to build safer structures. Yet, earthquakes do not harm people, who are outdoors away from structures[16].

So, Why are Earthquakes Necessary, or Useful?

Actually, earthquakes are the product of the same phenomenon that gives our landscapes their variety, texture, interest and beauty. Sediments are laid out in the ocean by river deposits and vertically compressed into sedimentary rock. The folding process described above molds the landscape into hills and valleys. The water that falls to earth and flows further carves the landscape into more complex shapes upon which vegetation grows to give us the rolling landscape beauty we have grown to favor over flat lands.

Mountain building is the most significant process that give us volcanoes and earthquakes. The stress also produces many of the minerals that are important to our daily lives, such as petroleum, many metals, such as iron, copper, aluminum, silver, etc. Important rocks, such as limestone, granite, marble, quartz and silica and a wide variety of materials that we extract from the ground, provide many of the resources fundamental to our technology[17].

Mountains contain rivers, which have been the primary source of water for many a civilization throughout human history. Rivers transport people, goods and a wide variety of essential products, as well as the drinking water we depend on for our survival. Where water originate too far from our settlements to be available to us where we live, it may be delivered by other source, such as precipitation, transported by the clouds. Water is also contained in most of the living organisms on Earth, many of which feed us regularly, (for example, fruit, vegetable, etc.). In the desert, succulent plants store water and the animals, which are natural stewards of their environment learn these water sources and pass on the knowledge to their offspring.

Emergency Preparedness:

Earthquakes are not easily predictable. When we rely only on human intellect to predict the occurrence of the next earthquake, our record has been unreliable. However, many animals have shown remarkable ability to be aware of approaching earthquake events before they occurred.

Considerable effective research has gone into understanding and predicting the behavior of animals. It is not unreasonable to conduct similar research into the behavior of certain animals prior to the occurrence of earthquakes. Such research may provide effective warning, from one's dog, cat or other animals, which may be available for observation. Being able, to interpret the unusual behavior of many of these animals prior to an earthquake, would provide useful warning of the coming event.

Such warnings were valuable to many people in traditional societies. Much of this knowledge is virtually unknown to us today, since we have come to ignore animals as potential teachers of our lifestyle.

Furthermore, the colonial legacy includes much purging of knowledge that was used by past traditional societies.

Nevertheless, it is important to be prepared for the eventual earthquake in regions susceptible to such activity. Incidentally, we know most of these regions, from seismic records. Moreover, as we will see below, an earthquake preparedness kit is useful for most other emergences that assail our everyday lives.

CHAPTER 9

— ❖ ❖ ❖ —

The Living Coastline

MANY PEOPLE TEND to see coastal land merely as desirable real estate. As such, the coastline is regarded as appropriate for permanent structures. In reality, coastal land is constantly and continually in motion. The coast is an active part of the Rock Cycle. It is also an integral part of the ocean's "sphere of influence".

The rock-type that makes up coastal land is usually dominated by sedimentary rock. The coast is exposed to a perpetually moving ocean, eroding, transporting and depositing sediment. A beach is essentially sediment in motion, very unstable in the presence of moving water. Any building placed on a beach is a sacrificial offering to the ocean, which is an agent of the perpetual recycling of the sediment on the ocean floor.

A cliff on the ocean front may be somewhat less unstable. It may not be as vulnerable as a potential building site as the beach. However, wave action occurs at the base of ocean-side cliffs, especially during storm events. Such wave action eventually triggers landslides on the cliff face. Thus, in the long run, the cliff and any retaining wall built to "stabilize" it are all a moving part of the ocean.

Most often, the land behind the cliff rises, leaving the cliff on the low edge of a slope. If that area behind the slope is cleared of vegetation, heavy rains can produce sufficiently generous volumes of drainage, which would further encourage erosion and landslides on the cliff face.

Examples of Coastal Movement:

The winter of 2004-2005 produced record rainfall in the Los Angeles basin and the adjacent Pacific coast. The summer of 2004 had seen numerous forest fires in the surrounding mountains. The following winter produced one of the rainiest on record. The result was a record number of disastrous flash floods and landslides.

The residents of Laguna Beach on the Pacific coast just south of Los Angeles had an unusual moving experience on the first day of June 2005. Most of an entire hillside in a subdivision experienced a major earthflow, (a massive landslide which would normally flow like thick molasses on a free hillside)[18].

The City of Laguna Beach is built on a steep slope of canyon-lands that reach down to an oceanfront of cliffs and beaches. This idyllic spot could be the seaside resort of anyone's dreams.

Its residents had survived the rainiest winter in recent memory less mudslides and landslides than some of its northern neighbors. The Laguna Beach residents thought the worse had been over. The excessive rain, however, had been quietly at work beneath the bedrock on which the houses had been constructed and, in particular, a layer of clay within the bedrock had trapped a volume of water, somewhat like a shallow artesian well. The separation it created with the surface rock, eventually triggered the earthflow beneath the buildings.

The moving land took blocks of houses and their streets with it, in an irregular pattern of seemingly random movements. Parts of streets were sheared off as sections of the hillside collapsed. House foundations were shifted, causing some houses to break apart.

Blocks of land carried houses with their street front downhill for several meters. Being "earthquake country", many residents at first thought it was an earthquake and took to the streets. The documentary spoke of people trying to drive to safety, only to find streets broken off in front of them. The panic was apocalyptic. It happened during the early morning, when many commuter parents and older children may have been leaving for work and school, while young mothers were fixing breakfast for toddlers.

This type of event, though more dramatic, is far less common than beach erosion. Yet, the disasters continue, because in modern civilization people simply do not pay attention to these events, or the global regularity with which they occur. For many they are regarded as a somewhat interesting human-interest story and assume "it would not happen to me". It may make the daily sensational news on T.V. and never be heard of again[19].

CHAPTER 10

❖ ❖ ❖

Floods

RUNNING WATER IS the most significant agent of erosion at the planet's surface. It is a dramatic interface between the rock cycle and the hydrologic cycle. On land, running water generally occupies rivers and streams. However, most of the water from normal precipitation seeps into the soil to feed underground drainage, which occurs out of sight.

Much of the drainage provided by visible streams is seasonal. In other words, many river and stream beds are dry for much of the year. In some instances, (as in deserts), a stream bed may be dry for years, occasionally accommodating the surprise flash flood. In the building of many cities and urban settlements, minor creeks and natural valleys are paved over and/or built around, leaving natural drainage courses virtually undetectable.

Flowing water follows the course of natural valleys, even when paved over, causing ground to collapse and landslides to occur in unexpected places. Such situations may sometimes complicate flood conditions.

A river carries many times the volume of sediment per gallon of water during flood stage compared with normal flow. In flash floods, the relative volume of sediment in motion is usually many times greater than in "normal" flood conditions.

Floods are the most common event disaster to strike human habitats. More people are killed by floods annually than all the other "natural disasters" combined. In addition, the origin of floods varies greatly:

1. They may follow forest fires.
2. They may occur in drought stricken regions, (especially deserts).
3. A water-storage dam may overflow, producing floods downstream.

4. To avoid a dam break at a flood-control dam, water may have to be released in controlled quantities, during an unusually large rain event, to avoid the dam being overwhelmed. This protection may create floods downstream, example: 1997-1998 floods in North-central California.
5. A slow-moving hurricane can release copious amounts of rain, sufficient to trigger enormous flood events. For example, many Bangladesh experiences in recent years have been caused by such an event.
6. Winter or springtime ice jams can cause epic floods.
7. Floodplains are the most vulnerable locations for regular flooding.

Floods and wildfires are essentially driven by weather phenomena. Often they are intimately related. In many regions a forest fire in the dry season may lay the foundation for excessive flooding in the next rainy season. An example is the Los Angeles basin in 2005, following excessive wildfires in the uplands during the fall of 2004.

Under unusually dry conditions, a forest fire may be ignited by lightning from a thunderstorm, (more often they are started by carelessness in the handling of fire, by a backfire from an automobile, or by arson). In each of the above cases, drought is usually the critical catalyst. Lightning can travel several miles from the actual storm. Sometimes, however, the drought is so intense that the rain from the thunderstorm may evaporate before it reaches the ground. Lightning strikes in a dry location can start a fire.

Desert Flash Floods: It may seem incongruous that floods would be facilitated by drought. Yet, severe drought can create ideal conditions for devastating flash floods. The critical combination includes:

- The absence of vegetation in a drainage basin, which accelerates the pace of runoff.
- The drought leaves bare soil susceptible to more rapid erosion from increase runoff.

- In flood stage, the power and scale of material carried by the flood water magnifies the impact of erosion.
- A sudden heavy downpour of rain, or a rapid snow-melt in the spring may provide the ideal discharge for a flash flood event.

The ideal place to provide this combination of such factors is a desert.

A deluge of three to five inches in the space of a few hours can devastate the homes in subdivisions built on land in a dry creek bed. A subdivision in St. George, Utah experienced a spectacular flashflood[20], when a dry creek bed turned into a raging river and washed away several of the houses abutting the stream and effectively changed the course of the river back to its original location.

The subdivision had been landscaped to relocate the natural stream bed at a strategic location to accommodate their golf course. The map was probably based on the topography of the landscaped subdivision. These altered landscapes often reroute streams to their desired location. It is not uncommon in the flood-stage of flashflood, for the river to re-take its original course.

The event of 2005 did have a positive impact on the community, however. Official maps could be updated to identify and indicate the original course of the river. They completed significant drainage improvement and creek restoration. Those improvements limited the impact of the 2012 flood event, which followed.

Dam Accidents account for some of the most devastating events. A dam failure is particularly disastrous, when it is not triggered by an unusual weather event. Such was the case, when a landslide fell into the Vajont Dam in Northern Italy. It sent a wall of water about 240 meters high down the narrow gorge of the Vajont Valley, which decimated five villages in the valley and killed some 2,500 people[21].

Slow-Moving Hurricane: A hurricane is a collection of thunderstorms. When an active hurricane slows down, its rainfall on the area through which it travels increases proportionately. This is likely to produce flood conditions. The hurricane deposits a considerable volume of rain per minute. By this formula, the longer the exposure to hurricane conditions, the more rain impacts the area[22].

Ice Jam Floods: During thaws, runoff from snow melt increases the flow of water in rivers. This increased flow raises the water level, which pushes up on ice sheets covering the top of the river. If the ice sheets break apart, they move downstream in a surge of large ice chunks. In places of restricted water flow (such as shallow bends in a river, or around low bridges), the ice can pile up and an ice jam will form. The jam may then buildup great enough to dam the river and cause flooding[23].

An even greater hazard often occurs, if the ice-dam suddenly breaks free, setting off a flash flood. It can be like a dam-break. As the water rushes downstream, it often includes blocks of floating ice, which add to the danger.

People who live near a river that freezes over in the winter should regularly check with the National Weather Service, which monitors such vulnerable rivers during the winter and early spring. If an ice jam is detected on one's river, local monitoring is essential, to include appropriate preparation to evacuate at short notice.

Above all it is *not safe* to walk on an ice jam. The ice is unstable and can shift without warning. Furthermore, if one falls into a river or lake beneath an ice jam, it is especially difficult for potential rescuers to find the victim, because he/she could be covered by many layers of moving slabs of ice.

The Public Has Short Memories:

The west side of St. George in July 15, 2012, was flooded by a heavy torrential outburst. So much rain fell that there was enough water standing in streets for children to enjoy swimming, as much as four feet deep on some streets. The area which incurred the most damage seems to have been on the riverside of the Santa Clara River settlement, not far from the 2005 now forgotten St. George's disaster[24],

A desert may average a few inches of rain per year. However, it does not rain every year. Instead, the more common pattern is a heavy rain event at about seven to ten year intervals). Usually the rain event is a major thunderstorm. Such an event provides the desert with its greatest episode of erosion and sediment movement. The irony is that the people most vulnerable to this surprise are the people who have moved to an urban subdivisions to enjoy the "good weather"[25].

For anyone, who has survived a major flood, the clean-up afterwards can be even more traumatic. Mold can build up rapidly. And, along with basic natural decay, the stench left behind can sometimes be overwhelming. That was a common complaint of people who returned to their homes in New Orleans after the destruction of Hurricane Katrina in 2005.

Living in the Flood Plain:

Floods are most common in low lying areas. Of all possible low lying areas, the flood plain can be expected to flood most regularly. So, why is it that people who live in the flood plain of rivers and streams are surprised by floods? And why are builder allowed to build permanent structures where floods can be guaranteed? These are some of the "planned disasters" that we build for in our modern urban environments, with impunity[26].

Hundreds of rivers of varying sizes have been studied, worldwide. It has been found that rivers with flood plains occupy the flood plain on an average of once every two and a half years. Engineers may interrupt that natural cycle by building levees, dams, dredging channels, altering the course of the river, or other adjustments. The long-term effects are varied, but most effects are bad and can sometimes be catastrophic. The Mississippi River has many cautionary tales to tell.

Taming the Wild Mississippi: Floodplains are strips of land along the main course of a meandering river in its mature stage that flood regularly. The natives, who lived along the Mississippi studied the landscape and passed down knowledge of its habits from generation to generation. They lived in the harmony with the river, settling on the high banks above the river and leaving floodplains undeveloped, except for the cultivation of crops.

When the wave of European settlers began to invade the North American interior, they were in a race to acquire whatever land they could occupy. Any land next to flowing water seemed to be ideal for growing crops, especially if it was flat. The natives were regarded as ignorant, primitive and without an understanding of their modern European lifestyle. It was also assumed that the natives had so much land that they were wasting it. Land on the floodplains was some of the most fertile for agriculture.

Their farmland was flooded often, causing considerable property damage and loss. The farmers complained to the government about the inconvenience of the constant flooding. The Army Corps of Engineers heard their cry and "declared war" on the Mississippi. They built up the natural levees to considerably greater heights to tame the "wild river" and their efforts had immediate success.

However, flood waters carry unusually large quantities of sediment, which used to be deposited on the other side of their natural, low lying levees. Each flood, under normal conditions, brings new deposits of nutrients to the floodplain, making it rich and productive farmland. The low-lying natural levees also allow the flood waters to drain back into the main river channel after the flood.

What the Army Corps did not realize was that stopping the floodwater from following its natural course, was causing the excess sediment to be deposited **in** the river, building up the river bed. Over the many years of stopping the floods, building up the levees, not only raised the river bed, but also steadily reducing the capacity of the river channel to handle the mighty river's discharge[27].

As the river traffic increased to accommodate the industrialization of towns on the river, the Army Corps had to take on the additional task of maintaining deeper channels to allow for larger vessels. Dredging the center of the river channels merely made the sides of the river channel shallower.

Furthermore, the Mississippi is no ordinary river. It drains almost half of the continental USA, which includes the part of the country that receives the largest amounts of rainfall, (more than a half of the rainfall). Unusually large rainfall years have produced sufficient discharge to over-top the levees. The solution they chose was to build up the levees still higher.

Higher levees merely postpones the occurrence of each subsequent flood. Each upgrade widens the main course of the river, so that subsequent floods carry increased volumes of water, inflicting more damage and depositing more sediment in the river channel. Insurance and other replacement costs increase and people demand more protection from the river.

An example of this recurring dilemma occurred in 1927, when an unusually heavy rain event in the region became disastrous. The main river channel had grown wider and its capacity had increased significantly, which was good for handling river traffic. By now, the floodplain was packed with farms and more heavily used than ever before. The levees had been built up with sandbags in many sections and the banks began to break. This flood was the worst in US history at the time. Thousands of people perished and hundreds of farms were devastated.

The flood of 1993 was more costly and did more damage than the 1927 event. More people were impacted, because the population along the Mississippi had increased over that of the earlier flood. One town, Valmeyer, had sustained so much damage from the flood, (90% of the homes had sustained significant damage), that residents voted to take the Federal Emergency Management Agency offer to relocate the entire town to higher ground not far away. The 500 acre site was located on a bluff overlooking the river. By Dec. 1996 some 450 of the 1600 displaced residents had settled into the new site. The 2000 Census recorded 1,141 residents in the town[28].

In recent years, the threat and the consequences to farmers is even greater. Insurance companies are looking for ways to avoid paying their higher costs, as the impact of global climate change is an ever increasing flood capacity[29].

CHAPTER 11

Hurricanes

IN THE MONSOON regions of the world, hurricanes are a common weather phenomenon during the summer. The region most susceptible to hurricanes is Southeast Asia, where they are called "typhoons". This weather phenomenon is also common in the Caribbean, Southeast USA and Central America, Australia and New Zealand and the Pacific Islands. In South Africa and southern South America the occurrence of hurricanes is far less common.

A hurricane is made up of several thunderstorms. They usually result when thunderstorms form in clusters over a warm open ocean. They organize themselves into an integrated storm mass, which is often called a "tropical depression".

This disturbance requires huge amounts of energy, which comes from the warm water in the ocean. Such energy fuels numerous updrafts, each creating a separate thunderstorm. Together, they produce a complex storm system, which fires up the hurricane's engine, sending the thunderstorms into rotation around a distinct low pressure center, called the "eye" of the hurricane – weather's greatest of rain maker.

<u>Anatomy of a Hurricane</u>: It is interesting to note that hurricanes are common in regions that also experience long periods of drought that may last for years. Thus, hurricanes arrive to heal the land of the drought symptoms. In 2011, 81% of the state of Texas experienced extreme drought during the hurricane season. In September Hurricane Lee came on shore in Louisiana, missing most of Texas, as it traveled northeast, eventually carrying heavy rain all the way to New England. Between 2nd and 10th September Lee precipitated a total of 29,000,000,000 gallons of water, according to a Weather Channel report[30].

Texas needed a hurricane at that time. However, if one were to hit the Texas drought-stricken region of the state, would there be sufficient tree cover to absorb most of the precipitation before it ran off into its streams? Would an impact like Hurricane Lee create flashfloods before sufficient water soaked the soil to heal the drought?

Prior to the arrival and domination of the monsoon regions by European society, the natives in the region maintained most settlements within the forests. The forest cover provides a natural defense against the invasion of strong winds and the storm surges, which accompany the hurricanes, as they move onshore. The forests absorb and neutralize the strong winds and the damaging surge of large waves whipped up by the hurricane winds travelling on the surface of the ocean. These storm surges can be driven inland by hurricanes.

The protective phenomenon was demonstrated by the forest culture of the Andaman Islands, off the east coast of India, during the 2004 tsunami. Much greater than a hurricane storm surges, the tsunami which killed thousands both east and west of these islands caused no deaths on the islands that were occupied exclusively by natives. They lived like their primitive ancestors in the forest. The only deaths were on the other islands cleared and developed by urban settlers[31].

In our experience, as we trace a developing hurricane on its journey across the Atlantic Ocean from the coast of West Africa toward the East coast of North America we cannot predict the exact day or time it would make landfall in North America, (or exactly where). We can only guess as to the route these tropical depressions may take across the Atlantic. However, when it is due to make landfall, our weather reports can give us two to three days warning.

We further complicate our decision-making process by expecting to know the exact date, time of day and location of a hurricane's landfall. To expect such precision is unrealistic. Our weather is the product of almost countless factors: Warm ocean water is necessary to brew a hurricane. The hurricane forms in the presence of very low local air pressure. Higher surface temperatures in the adjacent continental interior draw the hurricanes toward it. The power of the hurricane decreases when it makes landfall and is robbed of the fuel and energy provided by the warm ocean.

Hurricane Katrina, August 2005:

Hurricane Katrina was not a typical event, (but no hurricane is typical). Some patterns repeat themselves in almost every hurricane, however. In the North Atlantic a tropical depression first forms off the coast of West Africa. Such a depression generally travels across the ocean toward the Caribbean islands off the coast of southeastern North America in five to seven days. Roughly a half of these depressions gather strength and become a hurricane on that journey west.

Hurricane Katrina had an unlikely beginning as a small tropical depression on 13 August 2005. It dissipated the next day. However the ingredients for another articulated depression remained, hovering off the coast of West Africa, near Cape Verde Islands. It began to reorganize again on August 17th. On 23 Aug. it had gained strength sufficiently to be a tropical depression once more and began to move west at a faster rate. By Thursday 25 Aug. it had arrived off the coast of Florida as a category 1 hurricane[32].

Katrina slowed down and weakened as it crossed the peninsula, only to explode into action when is hit the Gulf of Mexico. During Friday 26th, Katrina went from a weak category 1 to a category 4 heading to New Orleans, with its eye growing in size from a width of less than 100 miles to over 200 miles wide on Saturday 27th. At that time the ring of thunderstorms around the hurricane was 400 miles wide. It seemed to slow down as it crashed into the Delta, with the eye a little to the East.

East of the eye of the hurricane, the winds are significantly stronger than those on the West. So, at landfall, New Orleans was spared the worst of the assault, which bore down on the Mississippi Coast on the early morning of Monday 29th. Emergency preparedness provisions were particularly critical at this point, especially for public safety and "homeland security" promises.

On Sunday 28 August 2011, in the wake of Hurricane Irene, ten rivers between Pennsylvania and Maine were at record flood levels. Prior to this hurricane event that year, New England had sustained record rain and snow amounts since early in the previous winter. The hurricane rain added to days of water-logged soil. In contrast, the Southeastern states had just ended a severe drought episode and their soils were sufficiently ready to absorb the hurricane downpours[33].

The primary impact of Irene on land was the excessive rain. The following year, Hurricane Sandy transported hurricane-force winds and a storm surge. The impact was greatest in New Jersey, New York and the New England coast[34].

Except for Australia and islands in the Southern Pacific hurricanes are seldom seen in the southern hemisphere. The Indian Ocean also sees occasional hurricanes. Because of the narrow land mass of sub-tropical South America does not attract hurricanes[35].

CHAPTER 12

❖ ❖ ❖

Tornadoes

As NATURAL EVENTS go, tornadoes are probably the most terrifying. Their scope of damage is generally small, when compared with a hurricane. However, their impact at the point of contact is usually more intense than a hurricane. For anyone who has lived through a tornado, life can never be the same.

A warning of an approaching tornado is usually quite accurate. The warning is issued only when it has been sighted. The alarm goes out in a specific area of a few square miles, around the spotting area. However, it comes with very little warning time. One often has merely minutes in which to respond to that official warning.

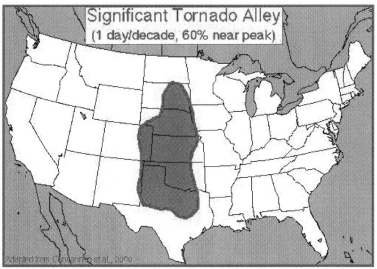

Map prepared by the National Oceanic and Atmospheric Association, (NOAA)

Modern weather reporting puts us closer to this meager level of prediction than anything previously available to us. From Doppler radar, weather experts can identify areas currently displaying the ingredients necessary to spawn a tornado. Wherever cold air coming off the Rocky Mountains meets warm moist winds from the Gulf of Mexico and Jet Stream flows above, the situation is ripe for tornadoes. In this case, a "tornado watch" is issued. Whenever an actual tornado is spotted, a "tornado warning" is immediately announced for the specific area around the spotting.

In regions where tornadoes are common, towns have sirens, which are sounded and people are regularly told what to do in such an emergency. Some cities have public storm shelters in high density locations, such as shopping centers and government buildings.

The Formation of a Tornado: A tornado grows out of a thunderstorm, or a super-cell. If the undercutting cold air is sufficiently cold, hail is possible, (sometimes golf ball or baseball-sized hall-stones occur and can smash car windows). Rain can be excessive. Indeed, the rain is often so heavy that it may obscure the rotating tornado cloud, which is the identifying image relied on to spot a tornado. If the tornado forms from a super cell, it may be so wide, (up to a mile wide), that tornado spotters may not realize that the blanket of rain they see actually contains a tornado.

On the ground, people have to be aware of developments around them. As a tornado approaches, certain places are particularly vulnerable. People living in mobile homes in a tornado prone region should have located a safe place to evacuate to, ahead of time, in the event of a local tornado warning. In many towns tornado evacuation buildings and/or storm shelters are available to people, who need a safe place to evacuate to. This includes anyone, who may be away from home, or those who may be on the street at the time the warning is issued[36].

Most traditional societies in humid areas have had a long tradition of "reading the clouds". A friend, who grew up in the Australian Outback tells of her family's aboriginal neighbors coming to her mother at the time of her birth to help with her delivery. Her father was at that moment pre-paring to wheel out their light airplane to fly them to the nearest hospital, hundreds of miles away for the delivery. The aboriginal elder hurried over and told him it was too dangerous to fly, because he knew a "willy-willy", (a tornado), would soon form in the area. Her father could not see any signs of such a development, but he respected the natives sufficiently to heed their warning. He stayed on the ground. A tornado did develop not long after he received the warning, along the path he would have taken to fly out. That evening, the women helped her mother deliver her.

Significance of the Jet Stream:

The Jet Stream is an upper atmosphere stream of air that flows roughly from west to east at about 200 mph. It may occasionally curving north or south, since it always flows at the junction between a warm and cold air mass. That junction determines the direction of the Jet Stream. Independent of the jet stream, westerly winds push any air mass with a low pressure centers across North America at varying mid-latitudes, (usually between 35 and 45 degrees latitude).

The Jet Stream operates primarily in the mid-latitudes, a region which is characterized by the prevalence of high and low pressure weather systems. High pressure systems blow winds outward from the center, while low pressure systems draw winds in toward the center. Low pressure systems are significantly more dynamic and active.

As the center of a low pressure system flows eastward, it draws in air from the Gulf of Mexico. The Gulf air, which is warmer travels northward in a circular pattern, which may seem to be coming from the southeast at first. As the system advances eastward the winds within the system would seem to shift to come from the southeast, from the south, then from the southwest.

The system also develops a warm front, which constitutes the warm air being drawn in from the Gulf. At the same time, a cold front will devel-op within the system, leading the cold mountain air from the Rockies. The warm air rises over the descending cold mountain air. This confrontation can often create stormy weather. The Weather Channel, a public service television station, makes current weather conditions available in the area of viewing. The chart below shows a path of yellow arrow heads, indicating warm winds approaching from the south. The blue line proceeding south from the low pressure center is the approaching cold front, with cold winds flowing east. Where they collide with the warm south winds, a tornado may develop. If the Jet Stream is travelling directly above, (white arrow heads in chart), such a thunderstorm, it may produce a tornado. A large and complex thunderstorm may develop into a "super cell", which can spawn several tornadoes over several miles of activity.

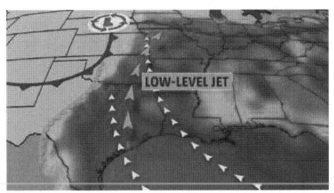

weather.com

7 Things You Should Never Forget When Tornadoes Strike by Sean Breslin, The Weather Channel[37]:

<u>1. Figure out a safe place to ride out the storm</u>

Do you live in a mobile home? Get out. Driving in a car? Get home as quickly as you can, and if that's not possible, get to a sturdy building.

<u>2. Get away from windows and get underground</u>

Regardless of where you're hunkering down, it should be as far away from windows as possible. Even if a tornado doesn't hit, wind or hail could shatter windows, and if you're nearby, you could get hurt.

You should make every attempt to get underground during a severe storm, either in a basement or storm shelter. If neither is possible, head to the innermost room or hallway on the lowest floor of your home. The goal is to put as many walls between yourself and the outside world. The image below, taken following the 2011 EF5 tornado in Joplin, Missouri, shows why this method could save your life. In many of those homes, the outer walls have been destroyed, but a few inner rooms are somewhat intact.

3. If a tornado appears while you're on the road ...

You should make every effort to find a safe building for shelter. If you can't find one, NEVER hide under an overpass. Instead, find a ditch, get down and cover your head. Get as far from your vehicle as you can to prevent it from being blown onto you.

4. Put on your shoes – and a helmet

If you're at home and severe weather is bearing down, prepare for the worst. If your house is damaged by a tornado, you could end up walking through debris that's riddled with nails, glass shards and splintered wood. The best way to ensure your shoes aren't scattered is to put on a pair before the storm comes.

If you own a bike helmet, be sure to put it on during a severe storm. It could save you from life-threatening head trauma if your home suffers a direct hit.

5. Keep your pets on a leash or in a carrier, and bring them with you

They're a part of the family, so make sure they go to a safe place with you. Make sure their collar is on for identification purposes, and keep them leashed if they're not in a crate. If your home is damaged by a tornado, it might not be familiar to them anymore, and they might get loose. Be sure to get them to a safe place or put them inside a crate while performing clean-up.

6. Don't leave your home and try to drive away from a tornado

If you made it home, stay there. Tornadoes can shift their path, and even if you think you're directly in the line of the storm, being inside shelter is safer than being inside a car. Traffic jams could keep you from getting out of the storm's path, or a small wobble could send the storm in a different direction.

7. Know your severe weather terms

- Severe thunderstorm watch: Conditions are conducive to the development of severe thunderstorms in and around the watch area. These storms produce hail of ¾ inch in diameter and/or wind gusts of at least 58 mph.
- Severe thunderstorm warning: Issued when a severe thunderstorm has been observed by spotters or indicated on radar, and is occurring or imminent in the warning area. These warnings usually last for a period of 30 to 60 minutes.
- Tornado watch: Conditions are favorable for the development of severe thunderstorms and multiple tornadoes in and around the watch area. People in the affected areas are encouraged to be vigilant in preparation for severe weather.
- Tornado warning: Spotters have sighted a tornado or one has been indicated on radar, and is occurring or imminent in the warning area. When a tornado warning has been issued, people in the affected area are strongly encouraged to take cover immediately.

A swarm of tornadoes occurs when several tornadoes come alive, roughly at the same time, in one general area. A "super outbreak" occurs when several tornadoes develop along a line of storms called bands.

The lines of thunderstorms may stretch across several states and may seem to follow each other as they moved across the country from west to east. On the Weather Channel these bands seem to march across the country, like soldiers in columns. For example, one band may stretch from Northern Texas, through Oklahoma, Kansas, Nebraska, Iowa, Minnesota, following another band from Arkansas, through Missouri, Illinois, Indiana. This band would, in turn, follow yet another band stretching northeast from Tennessee, through Kentucky, into Ohio. As that band continued eastward, it may stretch across eastern Tennessee and eastern Kentucky, into West Virginia and Pennsylvania.

One such band of storms in 2011 raged from 22 through 24 May in the Kansas – Oklahoma region. This band created several tornadoes in this way, behind a leading edge of severe thunderstorms.

Joplin, MO, just north of Springfield was worst hit on Sunday 22nd May, with 123 confirmed dead by Thursday, and still finding bodies in the debris. Hundreds of community folk and people from neighboring communities converged on the Joplin community.

Many dead were from St. Johns Medical Hospital in Joplin. One hospital wing was severely damaged. The volunteer team at the hospital evacuated the entire group of surviving patients in 90 minutes! Seven people were rescued alive in the debris in other parts of the city.

Radar picks up clouds of debris sucked into the vortex of a tornado. The tornado over Joplin was seen to draw debris 56,000 ft. into the atmosphere. Many areas experienced baseball-sized hail and exceptional rain out ahead of the tornado to add to the damage. Both the rescuers and other survivors risked being challenged by flooding as well.

A "tornado outbreak" is the group of tornadoes spawned by the same weather system. It usually includes all of the tornadoes in a contiguous-state region over a period in which there is no more than a 6-hour tornado-free gap, (including the horrific tornado that struck Joplin, Missouri on May 22 and outbreaks that brought 748 tornadoes in April, (including Tuscaloosa, Alabama). That set a record for any month, smashing the old record (542 in May, 2003) by more than 200 and obliterating the old April record (267 in 1974). April 27 was, to that point, the worst day of that record month[38].

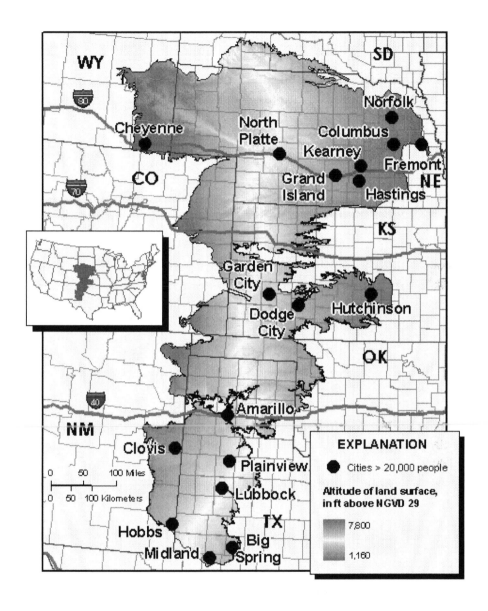

This region of "Tornado Alley" stretches from Texas to Minnesota. It would seem that tornado frequencies are increasing. In reality, there is no evidence that any more tornadoes occurred between the early 1900s and 1980s. The greater likelihood is that the rapid increase in populations and urban settlements provides many more targets in the way of travelling tornadoes.

The Joplin Emergency Planning team learned very valuable lessons from this disaster and put their experience to practical use as they set about planning for the next tornado. And, they were able to convince the federal authorities to build several strategically places community shelters, rather than try to equip each home with an individual shelter. They decided on this alternative, because many homes were not built with basements and many of the local residents lived in mobile homes[39].

The Drought Challenge:

Texas has had a significant draw-down impact on the Ogallala Aquifer. It is one of the largest aquifers in the Western Hemisphere and covers an area of 225,000 square miles. It stretches north to South Dakota, west into Wyoming, Colorado and New Mexico and east, almost to Iowa and Missouri, (see above map)..

Formed about one million years ago, it seems to have remained undisturbed until early in the 20th century. The water table is not a uniform distance from the surface, varying from approximately 50 feet to 300 feet and averaging 150 feet. The thickness of the aquifer varies from 200 in Texas to 1000 feet in Nebraska. Similarly, the demand for irrigation vary by state. Of the total withdrawal from the aquifer, 43% goes to Nebraska, 14% to Kansas and 10% to Texas.

The aquifer is recharged by rain and snow-melt. However, it has not kept pace with consumption. In recent years, the depletion rate has been increasing and many experts expect it to dry up in some areas[40].

CHAPTER 13

Forest Fires

FOREST FIRES ARE most likely to occur during extreme drought. That is why we most often look to the weather and monitor it closely during the "fire season" in many areas. Most forest fires are started by arson and/ or the careless use of open flames. But the weather is still the most critical factor facilitating and fueling the resulting fires.

Another unusual source of forest fires may come from lava flows in the vicinity of an active volcano[41]. Such fires can quickly get out of control, with extensive impact in remote areas that happen to be forested. Some are in grassland regions. Each event may be considered a blessing to some, or "damage" to others, depending on their respective desires or needs.

At a time, when suburban development is common in large metropolitan areas, many settlements are built into the forests. Often these forests are over-crowded, (they contain many times the number of trees per acre found in healthy forests). Thus, not only are these forests unhealthy; but the trees provide most abundant fuel for the next fire.

Forest fires are scary by any standard. Anyone who has been close to one would understand the level of fear they generate. Thus, the terror that gripped the growing suburban resident population between 1910 and 1990 is understandable. The popular response was to stop every forest fire as soon as possible. And, they became exceptionally good at it.

In partnership with ecology groups removing trees was discouraged in many neighborhoods, which was further promoted by the increased suburban spread into forested areas. However, many species of forest trees depend on the heat from a fire to split open their seed-carriers to release the seeds to germinate. The Giant Sequoia is one such example. A moderate fire would not be sufficient to split open its cones.

In addition, the hotter fire and its longer duration left a generous bed of ash on the forest floor, enabling the seeds to germinate easily after the subsequent rain. The new-found skills among forestry fire fighting became so efficient that many forest species, including the sequoias, were not receiving sufficient fire intensity to open and disperse their seeds. Thus, the future of some species was being threatened.

The moderate fires did leave a generous bed of fresh ash covering the land left bare and the subsequent rains would germinate whatever seeds were scattered. However, the species of trees, which required intense fire to expose their seeds would not be able to reproduce. Furthermore, the underbrush in these forests increased rapidly, with growing density, providing more generous fuel for future fires.

Mega-Fires:

The efficiency of fire suppression across the country and the influence of certain ecology groups created a new problem for the forests by 1990. Most of our forests had become over-crowded. Thus, many fires got out of control, especially in inaccessible area, given the often abundant fuel supply. Such fires get so hot, that they bake the topsoil, preventing future seeds from sprouting.

Such fires would consume so many acres of woodland that it often took a change in the weather to allow the fire fighters to eventually bring the blaze under control. These are considered "megafires".

The megafire creates its own low-pressure system, which would draw in dry winds from the surrounding drought-stricken areas. These winds would often attain hurricane-force speeds, severely fanning the flames further out of control. If such a fire reached an ocean, or a sufficiently large body of water, it would eventually attract cooler, moist air into the core of the blaze to reduce its intensity to a controllable level.

Many federal policies have already been put in place to reduce the overcrowding of large forests and their resulting impact. Many jurisdictions have reversed the moratoria on removing trees and have replaced it with considerable education to both local governments and residential property owners to reduce the density of trees on woodlots and natural forests, to safe densities. People are being taught to create defensible space around their homes so that, if a fire approaches, their home would be less vulnerable and much safer in forest fire conditions.

Homes in a coniferous forest region are especially vulnerable. Pine needles are very flammable. One of the worst forest fires in US history occurred during the Hayman fire in Colorado's coniferous forest belt in 2002, where entire neighborhoods were lost[42].

At different times, in a variety of conditions, a range of situations may develop in a natural forest to facilitate fires. Forests are formed by the climate in which they reside, as well as the topography of the land that supports them.

The Natural Forest:

Mixed forests are the norm. However, some climates produce a limited variety of quite similar types of vegetation. Two such "pure" forests are conifers and eucalyptus trees. Both are particularly highly flammable.

Polar climates produce mosses, lichens and other types of vegetation, which grow merely a few millimeters tall. In the mid-latitudes we find conifers and mixed forests, often depending on a small amount of annual precipitation, from a wet season, which comes seasonally during the calendar year. Tropical forests are naturally extremely dense. But, that density is due to abundant rain. Thus, the wet conditions discourage fires,

Rainfall amounts are probably the most significant determinant in Forest fire risk. Even in a humid tropical forest region forest fires can get close to, (and threaten) homes on occasion. Such events tend to occur during the dry season, where such weather conditions exist.

Another critical factor is topography. In hilly and mountainous regions fires are especially difficult to control. Fires travel faster going uphill than downhill. Thus, the homes "with the view" may be the most vulnerable.

Within many forests, soil types may become a significant factor influencing the size of trees at maturity. An outcrop of sandy or stony soil in a forest may cause the trees to be significantly smaller, or stunted. Sometimes a sufficiently large area of soil so affected can actually result in a break in the continuity of the forest cover, which may provide a fire break during a raging fire event.

In mountain regions, a high mountain could experience many different climates at different altitudes. In the tropics, for instance, a mountain that reaches up to 6,000 meters, (or over 18,000 feet), can have a sample of all the climates on earth on its slopes! Their tops are generally ice-capped and colder air has decreased capacity to hold moisture.

Moreover, parts of the forest higher on the slope may be drier and more vulnerable to fire risks. If fires denude the highlands, future rains can produce a further hazard[43].

CHAPTER 14

Less Predictable Natural Events

So far we have examined the more common natural events. Other natural events, though uncommon, can have tremendous impacts on human populations. Many of these impacts are significantly avoidable.

Tsunamis:

A tsunami, (or seismic sea wave), is created by an energy surge that travels across an ocean in the form of a fast moving wave. Traveling at speeds often in excess of 500 miles per hour, the waves resulting from such an energy surge are hardly noticeable in the open ocean. The surge develops into a gigantic wave when it approaches land, as the ocean depth decreases, causing them to increase in height as they approach land. They build up great height, because they are slowing down backing up on each other, carrying enormous weight and power!

The energy can be generated by any major disturbance on the ocean floor, such as an under-sea earthquake, or a landslide, or an explosive under water volcanic eruption, or other violent disturbances at sea. The telltale sign of such a wave is the sea suddenly receding outward from the shore to an unusual distance. Here, in order to build up the great height, these waves also draw back the water in front of them!

Soon, the huge wave approaches the shore at excessive speeds. The wave may become somewhat slower than its original speed out at sea, (as they slow down from 500 miles per hour). Anyone who lives near the ocean, or visits the ocean should be aware that a suddenly receding tide is a signal to flee for higher ground, immediately!

On 26 December 2004, the people along the coast of Sumatra did not understand that phenomenon and ran out following the receding tide to catch the stranded fish left in its wake. They were quickly overwhelmed by a tsunami wave, 20 meters high, which raced in a few minutes later. A tsunami travels so fast that no one could outrun it even given hundreds of meters head start from the receded tide.

Such events are the most uncommon of natural events that threaten our coastal habitats. Pacific coastal regions are the most vulnerable in North America[44].

An Indian anthropologist, who was studying the native people of the Andaman Islands, off the east coast of India, rushed over to see how the natives had fared in the aftermath of the tsunami of December 2004. When he got there, he found very little damage on the islands. One of the elders told him that the people had run into the forests when the sea went out, not a single person was lost[45].

A more reliable source of warning for upcoming events may come from the behavior of animals in our midst. They are sufficiently sensitive to subtle Earth-signs, because they are more in tune with Earth's rhythms than our clock and calendar instincts and our best science and technology. Even our pets could tell us a great deal more than we are capable of sensing. In effect, our pets can often help us detect valuable warnings, if we were more in tune with them and would only listen to their behavior.

The non-domesticated animals are generally better prepared to live with natural events than sophisticated humans and their domesticated pets. One report from Sumatra, the site of the greatest human loss in the aftermath of the tsunami tells of a team of elephants which bolted up a hillside before the arrival of the tsunami and saved the lives of the tourists they were carrying and the indignant elephant owners, who were obliged to follow the elephants, which they could not subdue. The elephant driver reported his experience of running up the hill behind the elephant and looking back to see the huge wave coming up the hill behind him, stopping and receding, not far from where he was, carrying with it all manner of debris[46].

Dam Breaks and Reservoirs Avalanches:

The landslide, which fell into the Vajont Dam in Northern Italy, (see Chapter 15 – Floods, above), sent a wall of water about 240 meters high down the narrow gorge of the Vajont Valley. "Megatsunami" is an informal term to describe a tsunami that has initial wave heights that are much larger than normal tsunamis. Unlike the usual tsunami, which originate from tectonic activity and the raising or lowering of the sea floor, these "megatsunami" have originated from large scale landslides or impact events in a lake. The Vajont Dam disaster was a man-made event, caused by faulty engineering in the construction of the dam[47].

A different, more horrendous hazard awaits the unsuspecting down-stream neighbors of dams built in developing countries, to store hazardous waste. These entrepreneurs exploit the ignorance of local governments unfamiliar with the complex requirements of safeguarding the public from the hazardous materials exposed by mining for precious metals. A recent dam disaster in Brazil portrays the nightmare in full color, (see photo below). The event was the collapse of a tailing dam released 50 million tons of iron ore waste, contained high levels of toxic heavy metals and other toxic chemicals in the river Doce, which 800 miles later entered the Atlantic Ocean, below[48].

Other Unusual Tsunami Experiences:

The landslide, which fell into the Vajont Dam in Northern Italy, (see Chapter 15 – Floods, above), sent a wall of water about 240 meters high down the narrow gorge of the Vajont Valley. That was like a "super-tsunami". It was a man-made event, caused by faulty engineering.

Evidence of unusual natural tsunami-type activity has been observed in isolated lakes. Sometimes, an avalanche can collapse its material into a lake to produce even larger "tsunamis", such as one lake in Northern Canada[49].

Brazilian Mine Disaster: United Nations Human Rights Press Release, by John Knox and Baskut Tuncat, 25 November 2015.

The mining waste spill impact over 6-weeks later along a 300-mile stretch of Rio Doce, still bright orange color from the spill (Vincent Bevins, The Times

In such an event, the limited, finite size of the lake causes the impact of the landslide to be amplified to create huge waves, sometimes to several hundred feet in height. One should be well aware of weather, while on a lake. A sudden wind can change a placid lake into a rough sea with large waves.

Unstable soil on hilly terrain makes those living on the land vulnerable to avalanches, soil creeps, landslides and other forms of earth movement that may collapse buildings and harm people. Following the unusually heavy rains in the Los Angeles area during the spring and early summer of 2005, major landslides occurred in various areas. Some events affected entire subdivisions[50].

Drought and Desertification:

Semi-arid regions are essentially regions in transition, between humid-tropical and desert regions. Transition regions are particularly vulnerable to the phenomenon of rapid climate change. Examples of such conditions are currently being played out in the Sahara and the Asian Steppe region. Regions stretching from Southwest USA into Northern Mexico, as well as, much of Western Australia, provide similar transitional conditions.

Given the seasonal fluctuations likely to repeat themselves, subsistent farming and plantation agriculture tend to be the preferred occupations, often including a nomadic existence. By its very nature, such living is not likely to be sustainable. Living on the edge of survival induces chronic poverty. Subsistent farmers are subject to crop failures. Subsistent herders, ranchers are often vulnerable to fluctuating prices in their respective markets.

Accordingly, the chronic subsistence often pushes those living under such an existence to the brink. The farmers and ranchers unwittingly over-consumed the local vegetation creating and sustaining drought conditions over an extended period, beyond which the land was not capable of recovering for decades.

The drought, itself accentuates the lack of adequate precipitation, when vegetation is not available to recycle the limited ground water. Thus, semi-arid conditions lead to desertification, which spreads. This was the case with the ranchers and herders of the Central Asian Steppe. They were enticed by the markets of Pacific-coast urban China, which developed high demand for livestock and furs from the Steppe. The resulting overgrazing of the land in the attempt by the local herders to meet these lucrative demands impacted the land's capacity to sustain the growing livestock populations[51].

Dust Storms:

Dust storms are natural atmospheric cycle events, not uncommonly occurring in arid areas void of vegetation. Because these regions are usually sparsely populated, these storms generally go without notice. However, as urban desert living becomes more popular among those seeking "good weather", an increasing number of suburbanites are being exposed to dust storms.

In recent years, huge dust storms have engulfed major metropolitan areas in Eastern China, Sydney, Australia, Phoenix and Tucson, Arizona, USA. These phenomena are all related to recent desertification of lands west of the affected areas, the result of unsustainable landuse. It is as if they brought their "dust bowl" to a place just beyond their suburbs.

Under these conditions, a gritty cloud of dust hangs ominously over the city, reducing visibility to levels sometimes unfit for normal highway traffic.

The Dust Bowl was the name given to an area of the Great Plains (southwestern Kansas, Oklahoma panhandle, Texas panhandle, northeastern New Mexico, and southeastern Colorado) that was devastated by nearly a decade of drought and soil erosion during the 1930s. The huge dust storms that ravaged the area destroyed crops and made living there untenable. Millions of people were forced to leave their homes, often searching for work in the West.

Many eye-witnesses of a sandstorm consider the experience their most terrifying ever! The top soil of marginal farmlands is sent airborne by prevailing winds with stormy velocity. Such was an incident in the

Phoenix, Arizona area. Speeding along a highway- suddenly engulfed by a dust storm- completely dark as night- visibility ended – pile-up on highway – dust storm followed by thunder storm and golf ball sized hail – fire breaks out within the pile-up passengers escape.

Sand Storms are Often a product of Desertification:

When the livestock consumed the native grass faster than it could be reproduced, the eventual drought conditions set in. Grass became scarce, livestock walked more miles in search of grazing land, they stirred up more dust on the bare ground, which continued to spread. Such a condition is a recipe for desertification.

The Steppe is now a fast expanding desert, contributing to the dust storms of East China metropolitan areas. A similar phenomenon is currently developing in Australia, where Sidney being visited regularly by sand storms. In both locations, such conditions were unknown two decades ago. Desertification is weather-related, with an assist from destructive land-use practices.

CHAPTER 15

— ❖ ❖ ❖ —

Essential Disaster Preparedness

SOME OF US prepare for inevitable disasters, while others merely build disasters into their happy-go-lucky lives. Many live with the expectation that emergencies occur in our everyday life experiences. Others live as if they are immune to the demands of emergencies. They expect life to be easy, with few interruptions, underlined by endless convenience. They expect modern technology to be able to solve all incidental problems.

In fact, as we have discussed earlier, much of modern planning and subsequent construction is undertaken, where the preparation for safety is often an incidental afterthought. In the normal pursuit of Earth stewardship, one is drawn into a lifestyle, which automatically familiarizes oneself with Earth's lifestyle and anticipates setbacks.

It is especially vital that people in this "market-driven" civilization be alerted to the significance of landforms, weather patterns and other visible natural dynamics that make up Earth's lifestyle. As we continue these discussions, an important theme to grasp is that we are often on our own and are required to look out for ourselves and our families in the face of many vulnerable situations that are overlooked by the administrators of urban development the world over[52].

Every winter, in the mid latitudes, dozens of drivers experience emergencies, which leave them stranded in their vehicle. In some cases, a vehicle may have slid off the road in an inconvenient isolated location. Others are caught in a blizzard, or may have run off the road in a sudden fog. Still others have an unforeseen mechanical breakdown, while driving in the desert. An occasional, but by no means impossible, situation is the careless driver, whose vehicle forces one off the road and/or into a collision with an unyielding other object.

Most Vulnerable Drivers:

The most vulnerable drivers are workers, whose job requires them to spend significant portions of their daily work driving. Service workers, delivery workers, utility service workers, messengers, mobile technicians, etc. are among the vulnerable. Emergency kits are especially important for anyone who usually transports children to games, sport practices, etc.

I worked for twelve years as a service representative for a regional utility and had the opportunity to rescue stranded drivers on a few such occasions. Because I drove my own personal vehicle on the job, my emergency kit was always with me. Thus, on family camping trips that kit came in handy, when faces with the emergency of a stranger along the road. My sons grew up with a sense of spontaneous preparedness for their own emergencies and the emergencies of others.

Earthquake Emergencies:

Earthquakes strong enough to damage buildings create emergency situations that are usually localized. Isolated buildings incur significant damage. People are trapped and have to be rescued.

Special Earthquake Preparedness kits are available in many chain drugstores in California. They are designed to provide the bare essentials one would need, if one were cut off for two days, without access to any outside assistance. The essentials include a basic first aid kit, drinking water for a family of four for two days, canned or dried food, etc.: (see emergency preparedness kit listed below).

However, under extreme conditions, such as tsunamis in Japan in March 2011, or Malaysia in December 2004, or Hurricane Katrina in New Orleans in 2005, extensive areas are devastated and virtually all local infrastructure are broken down. Electricity is unavailable, running water is scarce, shops are closed and commerce has ground to a halt.

Under such conditions, survivors are left to their own devices. As such, a survivor can manage with a basic emergency preparedness kits.

Many practical resources are available to us today, provided we take the time to explore the neighborhood in which you live. Go for walks in the neighborhood and see if you can identify the high points from the lowlands. Are there places that flood occasionally after heavy rain? How close do you live to those areas? How high above those areas do you live?

Make time to visit your local city hall, or town, or village center and ask the local planning department to show you their map of flood-prone areas. Get them to locate your house on that map. Ask them to show you the watershed area on that map, so that you can identify the watershed serving your neighborhood. Does the watershed include a large area beyond your town or city? Is your home close to a large river? Remember that floods kill more people each year than all the other "natural disasters" combined! So, get to know your watershed area.

Always keep an emergency survival kit in your home, (in addition to the kit in your car). Keep the home emergency kit in an easily accessible location, well known to everyone in the household, especially the children. Take it out at least once a year, (how about Christmas time!) and check all the equipment and replace whatever is no longer functioning. Review with the children the role and function of the kit to help them to know what to do in an emergency, and how to use each critical item, (see emergency preparedness discussion above).

Observe what other natural events are common in your region and take steps to adapt your emergency kit to include preparedness for such events. Observe how local residents regard each natural event. Awareness of local attitudes to emergencies will help improve your preparation for such events. A critical part of effective preparation is knowing how seriously neighbors take these natural events.

Not only are people often left to their own devices after many disasters; whatever insurance coverage they may have for damage and loss may not be sufficient. Indeed, displacement from one's home and the death of family members, friends and neighbors can be almost irreversible to many victims and relief may often be years in coming, if at all. Consider how many of the victims of Hurricane Katrina, (2005), were still unsettled in 2010.

For all the levels of vulnerability that exist in community and in travel, the idea of having access to an emergency kit at all times makes sense. This is particularly critical for parents with children.

Emergency Preparedness Kit:

Safety Checklist: (for a home during a State of Emergency – assume being three days without normal urban communications, power, etc.)

Water: store 3 gallons of water per family member.

Flashlights: with extra batteries, plus one that runs without batteries.

Radio: with extra batteries, or one that runs without batteries.

Canned soups, fruit and vegetables, (plus can opener)

Healthy snacks, including energy bars.

Clothing: 3 outfits per family member.

Personal and bathroom items, (such as: glasses, toothbrush, soap)

Disposable breathing masks and goggles.

Matches: in water-proof container!

Medicine: a full supply of basic family needs.

Whistles: one per family member.

First Aid kit: to include tape, bandages, small scissors, bandaids.

Basic traveling tool kit: wrench, pliers, screw driver, hammer.

Photos of family members, label with name, date of birth, date photo was taken, height, weight, medications, address and phone number.)
Important papers: (copies of birth certificates, health, finance and insurance documents
Emergency contact information: 4 contacts, include info for someone out of state.
Pet supplies
Emergency blankets
Emergency ponchos
Canvas gloves
Sturdy shoes
Money
Family Disaster Plan
Map of your city or neighborhood
Fire Extinguisher
Smoke alarm and carbon monoxide detector

Store as much of this as possible in carrying containers, placed in a convenient location that all family members are familiar with, like a coat closet near the front door. If you have to evacuate your home by car, take this kit with you. Or, better yet, your automobile should contain such a kit as well as the home, since, more emergencies occur while driving than at home in most cases.

CHAPTER 16

❖ ❖ ❖

Listening to the Land

THE NATIVES OF traditional societies throughout North America knew Earth's lifestyle sufficiently well to participate in the harmony of all things natural for tens of thousands of years. They lived along the Atlantic coast and the Gulf of Mexico, areas susceptible to hurricanes. And they chose to live *in* the forest, rather than clear the forest to build permanent structures. They appreciated the fact that the forests protected them from the strong winds of the hurricanes and the storm surges, which usually followed.

Had the Europeans been able to appreciate the wisdom of their hosts and observed their lifestyle with appropriate respect, they would have learned the hurricane phenomena. Based on that knowledge they would probably have developed the coast differently. However, given the state of social life in Europe at the time and the mandate of the Doctrine of Discovery, the European effort at colonialism was destined to abuse Christianity and use it as a weapon to give religious justification to the widespread plunder of the Americas, Africa and the rest of the colonial world, 53.

The key to living in harmony with Earth is our willingness to be far more observant, more patient, more compassionate and humble than is customary in modern society. We must dismiss the well-learned attitude of superiority over animals and inanimate objects, including Earth, itself. The "wild" animals are quite well attuned to Earth's lifestyle. As such, they freely share their observations with us. We just have to observe and learn from their lifestyle. They give us adequate warning and invitations to participate in each upcoming event.

Such participation is necessary, with respect to the Mississippi and its flood plains. Had we been thus involved in the life of the river, we could have saved many thousands of lives and avoided many of our choices in landuse along vulnerable rivers and streams. Coastlines would probably have been preserved as public easements, spared development, respecting predictable ocean invasion and beach erosion.

Other lessons to have learned from the wisdom of the land and its people include the following:

- Clear-cut forestry in mountainous regions would result in excessive runoff after heavy rains, contributing to mud slides, flash floods that devastate settlements downstream.
- Suppressing natural forest fires lead to over-crowded forests with ample fuel for mega-fires over time.
- Not all landscapes are conducive to urban development.
- Many climates are not ideal for urban-scale settlement.
- Natural forces protect landscapes.

Before the European settlers moved onto the plains and built farmsteads, the grassland habitat was virtually undisturbed. Trees were unusual in the flat grassland surface. They only existed along rivers and streams. The farmstead, itself was an isolated elevation on the plain. Even then, as continued to be the case on today's farms in remote areas, very little damage is incurred from tornadoes.

Villages, towns and cities provide a more vulnerable target for tornadoes on the plains. Furthermore, far more damage is possible, as more settlements are built in Tornado Alley. In 2011, one of the worst disasters occurred, when Joplin, Missouri, a city of over 50,000 had about 30% damages, 116 killed and 2,000 buildings damages.

Today large metropolitan areas are located in tornado regions, such as Dallas, TX, St. Louis, MO, Minneapolis, MN, Oklahoma City, OK, etc. The question is no longer: "If" they will be hit by tornadoes; but "when".

Planning for Disaster:

In our relentless effort to grow, we stubbornly build wherever real estate is available and utilities can be provided. We improve our defenses against "natural disasters". However, our most noteworthy growth continues to be in tornado casualties.

The public attitude continues to describe the weather as "good", if there is warmth and sunshine. Cold and rain are considered "bad weather". "Mother Nature" is a force to be feared as a dangerous, unpredictable threat to our happiness. In our "war against nature", comments such as: "the wrath of Mother Nature" and "angry storms" are often used to describe such natural events as: earthquakes, volcanoes, stormy weather, tornadoes, etc. Weather reporters often complement such prevailing attitudes.

Such attitudes make it difficult for our civilization to appreciate the true value of Earth and the harmony with which all events complement each other to support Earth's lifestyle. Infatuated with the scope of modern technology, we continually measure our achievements by the extent to which the technology attains control of natural events on the planet.

In fact, we control nothing significant in Earth's realm. We merely "run interference" along the path of natural events! Moreover, in our efforts to control the environment, we have contaminated so many natural processes and so disrupted the balance of gasses in the atmosphere, that the global climate has been altered sufficiently to create conditions previously unknown in recorded human history[53].

Why do These Events Become Disasters? In modern society, we actively ignore our physical environment. We approach nature as the enemy to overcome, which sets us up for failure. Nature is neutral, with its own disposition to serve all life on the planet.

Most natural events follow quite predictable patterns. The National Weather Service introduces the weather as a force in control of what weather exists. Traditional societies observed weather patterns closely and based most of their ongoing activities on such observations as weather changes, the seasons and many other related natural phenomena. Their genuine humility left them better prepared, (than we are today), for the natural events of their respective regions.

The people in traditional societies were well tuned into the behavior of their local animal neighbors. Thus, they could take cues from the behavior of animals', (and plants), to anticipate what event was probably imminent. So, they could take necessary evasive action accordingly. For instance, the sudden absence of birds warns of an approaching storm that may not otherwise be obvious to a casual observer.

Other observations of the surroundings were critical. They developed an intimate knowledge of their immediate physical environment. The animals and plants in their neighborhoods were familiar to all. Where they chose to live reflected such knowledge and awareness.

We need to overcome our history of neglect, environmental abuse, indifference to the natural world and our selfish concept of freedom as entitlement to privilege without responsibility for the impact of our actions on the environment.

All these attitudes conspire to do us in. Our accumulated abuse of the planet has resulted in a climate so extreme that we can no longer expect to predict with a measure of accuracy we used to count on, what our future may hold.

This present climate is completely unknown to human experience[54]. Against the backdrop of the pride with which we assume modern technology to be capable of subduing and controlling Earth's function, our lifestyle continues to lead us to disaster.

Notes

1. See discussion on: Canine Telepathy: "Can your Dog Read your mind?", by Catherine Pearson, Huffington Post, August 12, 2011.

2. Pets will warn us of earthquakes, (and tsunamis if we live near the coast), many hours before the earthquake strikes. Plants continually

tell us even a great deal about the local weather and climate. The shape of a tree can often tell us the direction of the prevailing winds, which shaped it. A broken branch may suggest that it has survived many a severe storm. A core sample of a tree will reveal the age of the tree, by counting the annual rings. The pattern of its annual rings in an old tree may report drought patterns over the past, up to a thousand years, or more. By observing our surroundings and reading the landscape, we can tell if the home we are choosing to live in is located in a floodplain, or in a natural valley vulnerable to future flooding.

3. The Adam and Eve story in the Christian story of creation, taken in the context of the original Hebrew tradition calls on the faithful to care for creation. The idea of subduing nature flows more naturally from our modern lifestyle, which is largely oblivious to Earth's lifestyle.

4. See: Crater Lake Story, National Park Collection, Harper Ferry Historical Association, (a video tale of a native legend about the formation of Crater Lake.

5. An account of the lahar, which engulfed Armero, (1985), is available: www.google.com Enter: "How Volcanoes Work: the Nevado del Ruiz eruption".

6. For a description of volcanic fissures, see www.google.com Enter: "Nyiragongo-fissures. Also see www.pbs.org/wgbh/nova/transcripts/3215

7. Lake Nyos Story is reported by Google. See www.google.com Enter: Lake Nyos Story. [p. 22]

8. U.S. Geological Survey, Fact Sheet 113-97: "The Cataclysmic 1991 Eruption of Mount Pinatubo, Philippines, (report on the impact of

subsequent lahars.) Although Mt. Rainier is not susceptible to such large monsoon downpours, so large is the existing ice-pack, that a sudden melting ignited by a spontaneous buildup of magma heat can produce equivalent water runoff.

9. Ibid., Tarbuck and Lutgens, 11th ed. pp. 266-267

10. Nova documentary: "Mystery of the Megavolcano", aired on PBS 15 February 2005, describes a process used to identify pre-historic volcanoes in their respective locations. In the report, five scientists from different disciplines, working independently, are brought together by a mysterious volcanic ash sample to discover the existence of a caldera, known to the world as Lake Toba, Indonesia. The lake is about 1,080 square miles and contains a large island in the middle, about a third of the area of the lake.

11. Acharyya, S.K.; Basu, P.K. (1993). "Toba ash on the South Asia and its implications for correlation of late pleistocene alluvium". Quaternary Research 40: 10–19. doi:10.1006/qres.1993.1051.] [Scrivenor, J.B. 1931. The Geology of Malaya (London: MacMillan), noted by Weber.]

12. Huang, C.Y.; Zhao, M.X.; Wang, C.C.; Wei, G.J. (2001). "Cooling of the South China Sea by the Toba Eruption and correlation with other climate proxies ~71,000 years ago". Geophysical Research Letters 28 (20): 3915–3918.

13. Quote by Dr. Pat Abbott, Professor of Geology, San Diego State University, Host; DVD, Earthquake Country, Los Angeles, product of Instructional Technology Services, San Diego State University.

14. Pillow lava results from volcanic action on the ocean floor. When magma emerges into the ocean water from the eruption in the

ocean, the outer layer solidifies and quickly forms a shell around the still-molten mass. As the deposits gather on the sea bed, t takes on the appearance of black sandbags, piled upon one another. Later, as the sea-bed is folded, if such deposits are located on the upward fold, it can eventually become attached to a slope in a coastal range of hills, or an offshore island hillside the shape is characteristic of lava formed in the ocean.

15. The source of each earthquake is called its focus. Since this location is underground, one needs to have a reference point at Earth's surface. The earthquake discharges two distinctly contrasting shock waves, ("P" waves and "S" waves). They travel through the Earth at predictably different speeds, making it possible to calculate the distance from which they travelled. Today, thousands of seismic stations exist in the vicinity of any point in regions susceptible to earthquakes. Thus, the combination of reports from any three seismic stations will serve to locate the focus. The epicenter is the surface point directly above the focus, see Tarbuck, 11th ed., pp. 194-106.

16. Ibid., San Diego State University DVD, <u>Earthquake Country, Los Angeles</u>,

17. One of the ongoing roles of the rock cycle is its ability to cycle minerals from deep within the planet, (hundreds of miles deep), to be accessible to us on, or near Earth's surface, (e.g. gold, diamonds, rare metals). Earthquakes and volcanoes facilitate such movement.

18. An in-depth description of a variety of slide phenomena created by saturated topsoil on hillside landscapes, Ibid Tarbuck, 11th Ed., pp. 102-111. In densely populated urban areas, such slides may take thousands of lives. In December 1999, one such slide took an estimated 19,000 lives, (see photo on p. 110 of Tarbuck).

19. Unfortunately most people see disasters as something that happens to someone else. In addition, people tend to ignore the events that are common in their neighborhood. However, we travel and may be in an unknown area at the time of one of their hazardous events. We should acquire a more vivid sense of how these events play out. From this vantage point, a related discussing on safety and disaster-preparedness is included in Chapter 18.

20. In a report was entitled: "Southern Utah floods prove hazard maps faulty", by Nancy Perkins, of the Deseret News, (Tuesday Feb. 8, 2005): "Flood plain maps include the flood way, or the predicted path of the flood based on engineering measurements and past experience, and the flood fringe, where the water will likely move based on topography and development outside of the flood way. Thus, one resident was surprised to find the stream, which had been 100 meters away was suddenly flowing through her backyard and taking out half of her house.

21. The details of this event were extracted from a study: The Vajont Landslide by Dr. David Petley, Wilson Professor of Hazard and Risk, Department of Geography, Durham University, United Kingdom. He has since compiled a blog, detailing the sequence of events that led up to the eventual landslide, which collapsed into the lake.

22. Hurricane Mitch, which stalled over Nicaragua and Honduras for several days, including one 3-day stretch, during which five feet of rain fell in the highlands. Hurricane Mitch remained a hurricane from October 26, 1998 to Nov. 5.

23. USATODAY.com – Ice Jams Can Cause Floods: Scott Kroczynski, National Weather Service, Copyright 2008 USA Today, a division of Gannett Co. Inc.

24. In 7½ years the floods of Winter 2005 had been forgotten. Amy Joi O'Donoghue, Deseret News on Sunday, July 15 2012 wrote: "St. George awash with flash floods from violent storms", a report in which she quotes the Assistant City Manager as saying: "Never in my 15 years have I seen water come down that much, that quickly".

25. Washington County residents found out in early January 2005 that the Santa Clara and Virgin Rivers were capable of ignoring those established flood routes and destroying homes, barns, orchards and anything else in the way. These altered landscapes often reroute streams. However, in the scheme of things in Earth's routine, such alterations are temporary.

26. As reported by Nancy Perkins, of the Deseret News, several days later, Tuesday Feb. 8, 2005. The report was entitled: "Southern Utah floods prove hazard maps faulty". The map was probably based on the topography of the landscaped subdivision: "Flood plain maps include the flood way, or the predicted path of the flood based on engineering measurements and past experience, and the flood fringe, where the water will likely move based on topography and development outside of the flood way." It is not uncommon in a flood of this magnitude, for the river to re-take its natural course.

27. The above Deseret News report further revealed that: "Losses due to flooding are not covered under typical homeowners and business insurance policies," Jensen said. "When floods destroy homes, owners are left without a place to live but are still obligated to pay off their mortgages." After the devastation of Hurricane Katrina, in 2005, flood insurance became increasingly difficult to purchase in many parts of the USA.

28. In a recent article by American Rivers, http://www.americanriv-ers.org A study on the impact of governmental and community initiated structural flood control efforts, entitled "21st Century Solutions to Protect People and River Health", they have found that: "Unfortunately, for many decades communities have turned to structural flood control, like dams and levees, that have destroyed the floodplains and wetlands that provide natural, free flood protection for communities. We need smarter, 21st century flood protection solutions that work with nature, not against it."

29. Structural flood protection provides a false sense of security about living in the floodplain, which can lure people into high-risk areas. When rivers rise, as they inevitably will, those within the floodplain can quickly become victims, according to: American Rivers report Dec. 2012, 1101 – 14th St. NW, Washington, D.C. 20005]

30. Hurricane Irene, 2011, devastated much of the Eastern Seaboard, northward from North Carolina, 8/26/2011 to Canada, when it hit on 8/28/2011.

31. When they had a chance to consider what happened in the Andaman Islands in the face of the tsunami, the analyst was able to appreciate that the natives on the isolated island had learned from the ancestors, what those who had moved to town had not learned from their modern educators. Thursday, 20 January, 2005, "Tsunami folklore 'saved islanders'" by Subir Bhaumik, BBC News, Andaman and Nicobar Islands.

32. The intensity of each hurricane is driven by the level to which the atmospheric pressure falls in the eye of the storm. The Saffir-Simpson Hurricane Wind Scale is designed to measure the actual speed of sustained winds on a scale of 1 to 5. This scale also estimates potential property damage. Hurricanes reaching Category 3 and higher

are considered major hurricanes because of their potential for significant loss of life and property damage. The scale is (1) 74-95 miles per hour, (mph); (2) 96-110 mph; (3) 111-129 mph; (4) 130-155 mph; (5) over 156 mph: NOAA/National Weather Service, Miami, Florida 33165, nhcwebmaster@noaa.gov

33. Hurricane Irene, 2011, devastated much of the Eastern Seaboard, from North Carolina, 8/26/2011 to 8/28/2011, when it hit Canada.

34. Several documentaries reporting Hurricane Sandy in 2012 show the extent to which the Hurricane Katrina awakened an entire population to better emergency preparedness. See: "After Hurricane Sandy, Helping Hands" Also Expose a New York Divide – by Sarah Maslin NIR, NYTimes.com @import - November 16, 2012

35. The World Meteorological Organization of Arizona State University, School of Geographical Sciences, [P.O. Box 870104, Tempe, Arizona 85287-0104] presented a study: "First Identified South Atlantic Hurricane". The study called it: "the first recorded hurricane in the South Atlantic basin since geostationary satellite records began in 1966. This storm came ashore along the coast of Brazil at Santa Catarina on March 28th with sustained winds of 65-70 kts (75-80 mph) and gusts of 85 kts (95 mph)."

36. Many major cities and towns have constructed emergency shelters at strategic locations specifically designed to house people on the street, or those who may not have a place to evacuate to during an emergency. These include out-of-town visitors and those who are too far from their homes to reach during an emergency. Such emergency shelters are more common in tornado prone regions in the USA.

37. "7 Things You Should Never Forget When Tornadoes Strike", by Sean Breslin, The Weather Channel, http://www.weather.com May 8, 2015.

38. "One of the main difficulties with tornado records is that a tornado, or evidence of a tornado must have been observed", to be recorded. National Climatic Data Center, National Oceanic and Atmospheric Administration, 2012.

39. Reported by Michael Bettes, The Weather Channel, Thursday 16 February 2012.

40. In the midst of Tornado Alley, an underlying drought persists. Yet, the farming practices are wasteful. They have been overdrawing water from one of the largest aquifer in the world. Some experts calculate that it will be exhausted in five years. A Drying Shame: "With the Ogallala Aquifer in Peril, the Days of Irrigation for Western Kansas Seem Numbered", by Lindsay Wise, The Kansas City Star, July 24, 2015.

41. Forest fires and other outdoor conflagrations are a constant threat to properties in the vicinity of the Hawaiian lava flows: See: Kilauea eruption updates (Jan 2013 - Oct 2014), Natural Resource Report NPS/NRPC/GRD/NRR—2009/163

42. A story from one of the worst forest fires in US history comes from the Hayman fire in Colorado in 2002. One resident was able to clear 150 bags of pine needles from around his house just weeks before that megafire struck. The entire neighborhood was evacuated and when he returned to his home he found his house intact, with minor damage, while seven houses around his were burned to the ground. 150 bags of pine needles is a lot of fuel for any fire!

43. 2004 was a year of many summer forest fires in the Los Angeles region. It left several square miles of uplands deforested, contributing to vast unstable acreage at the start of the rainy season. Winter is the rainy season in the southern Pacific. Thus, the 2004-2005

winter saw record rains in the Los Angeles and Southern California. The tree-less highlands offered excessive runoff to the rainfall. The steady, heavy rain doubled the runoff in some areas. Many canyons in the region experienced significant flashfloods throughout 2005. Also: see emergency preparedness discussion in Chapter 18.

44. Tectonic plate movement off the North American Pacific Coast are a continuous potential capable of spawning a tsunami. The Pacific Plate is subducting beneath the continental plate of North America and it due for an off-coast earthquake.

45. When they had a chance to consider what happened in the Andaman Islands in the face of the tsunami, the analyst was able to appreciate what the community, who had never left the forest had learned from their ancestors. They retreated into the forest and were all safe. In contrast, those who had become urban dwellers had lost contact with their ancestors' customs when they moved to town. Many of them lost their lives in the tsunami. Thursday, 20 January, 2005, "Tsunami folklore 'saved islanders'" by Subir Bhaumik, BBC News, Andaman and Nicobar Islands.

46. Soon after the 2004 Tsunami in Southeast Asia, a story surfaced about an elephant that bolted up a hill minutes before the tsunami struck the coast where the elephant and his owner were standing. The owner's life was saved by chasing after the disobedient elephant. See transcript of the video news report: "Senses Help Animals Survive the Tsunami", By Charles Sabine, Correspondent NBC News updated 1/6/2005 7:54:09 PM ET2005-01-07T00:54:09

47. Refer back to Chapter 11 on floods, especially section on "Dam Breaks". The details of this event were extracted from a study: The Vajont Landslide by Dr. David Petley, Wilson Professor of Hazard

and Risk, Department of Geography, Durham University, United Kingdom.

48. After an 800 mile trip of contaminated spillage from the BHP Billiton Mine, the sludge tailings enter the Atlantic Ocean. See details in report: Brazilian Mine Disaster: United Nations Human Rights Press Release, by John Knox and Baskut Tuncat, 25 November 2015. See also: "Brazilian Mine Disaster Releases Dangerous Metals", by Luisa Massarani, 21 November 2015.

49. A paper written by C.M. Riley describes a variety of situations, which could result in tsunamis in lakes, with devastating results: "Debris Avalanches, Landslides, and Tsunamis, 1980 www.geo.mtu.edu/volcanoes/hazards/primer/move.html

50. Bluebird Canyon, within the City of Laguna Beach, experienced a major landslide on Wed. June 1st, 2005. The landslide completely destroyed 17 houses and damaged many others. The entire hillside collapsed, as water from the excessive rain separated an upper layer of rocks, which slid down the hill, taking sections of streets with the houses. In all, over 1,000 were rendered homeless by the damage. The ordeal was aired in a video documentary on Storm Stories, Weather Channel.

51. Similar to the conditions, which precipitated into the "Dust Bowl" of Southwest USA in the 1930s, the story written by Jonathan Watts, The Guardian UK Rapid, Wednesday 20 May 2009. In his article: "Chinese Plan to Relocate 150 Million Eco Refugees" he described the effort of locals to reverse the original desertification in some regions of China, where farmers frantically planted trees in an effort to save their cropland.

52. In this respect, every driver should keep an emergency preparedness kit in his/her automobile. One never knows when one may be

stranded on a remote road, without convenient access to a rescue vehicle.

53. Pope Nicholas V issued the papal bull.: Dum Diversus on 18 June 1452. It authorized Alfonso V of Portugal to reduce "Saracens, (Muslims), and pagans and any other "unbelievers" to perpetual slavery. This facilitated the Portugese Slave Trade from West Africa. As a follow up to Dum Diversus, it extended to all the Catholic nations of Europe dominion over discovered lands during the Age of Discovery. These orders gave religious fervor to a colonial legacy, which came to be the standard for the treatment of indigenous natives of the lands later conquered by Europeans and the ruthless exploitation of their lands. Thus was born the indiscriminate environmental abuse of the industrial revolution.

54. Bill McKibben, in his recent book: Eaarth, "Making a life on a tough new planet", c2010, 2011, Makes the point convincingly, that new developments in climate change leave us without a precedent by which we may predict the consequences.

III

The Stewardship of Earth's Abundance

Stewardship is a Way of Life, Which Allows Us to Support Each Other in the Spirit of Sustainable Living.

PRACTISING STEWARDSHIP IN a community ignites a spirit of cooperation that becomes contagious. We see this phenomenon emerge in the aftermath of disasters of any kind, locally, as well as internationally. When a community experiences disasters, (from a multi-vehicle pileup on a busy highway in a sudden dense fog to a hurricane, flood, or a tornado), people simply come out to help, rescue, comfort, or otherwise assist those in need.

We are spiritually connected to each other. Genuine humanity attracts us to each other and that attraction often comes with compassion. Competition is also natural. In human interaction there is room for both.

In any community conflicts will arise. The traditional village valued a standard of neighborliness, which placed great value in promoting and supporting a spirit of mutual goodwill. Conflicts often required the village elders to deliberate on the issues involved. With an emphasis on harmony in community, their goal was generally to achieve consensus among themselves.

The colonial legacy, with which modern society is infused, has instituted a level of spontaneous competitiveness, which often seeks to discourage cooperation. Conflicts were decided by a vote. The primary goal was to establish winners and losers. This approach has been associated with democracy, but directly contradicts the spirit of cooperation necessary to cultivate programs and activities designed to serve the common good[1].

The colonial urban-industrial lifestyle has taught us to value a concept of individualism, exclusive of community. Yet, urbanization relies very heavily on the interdependence of millions of individuals working to serve each other in ways that support our global economy.

In other words, whether the emphasis is on cooperation or competition, markets still rely on exchange. However, in one setting, those who cooperated openly reaped the satisfaction that their labor produced a joint product in which they could see. They would also share in the value of their product. In the competitive setting there is little sense of teamwork. Whatever compensation they received would not be associated with their actual labor and only a small portion of the actual value of their labor would be compensated.

The spirit of Earth stewardship has a cooperative edge, which brings people together. Thus, in community it may become a common practice. Groups of community garden farmers in the inner city often attest to that sense of community it fosters where most of the other aspects of their neighborhood may not be as fulfilling.

The Challenge of the Current Climate Change Crisis: This study discusses the universal practice of Earth stewardship as an appropriate response to offset the current threat of rapid global warming. Section II outlined the practice of basic Earth Stewardship as a potential opportunity to live in harmony with Earth's lifestyle. Section I described the harmony and infinite sustainability of Earth's lifestyle.

In Section III, we begin to explore what stewardship would look like in today's modern setting. Thus, we begin by exploring the nature of the climate crisis and what in our modern lifestyle has precipitated the crisis.

The Climate Crisis: Earth's atmosphere is made up of a specific proportion of nitrogen, oxygen, argon and a trace of carbon dioxide. Because the atmosphere is also the primary vehicle designed to extract fresh water from its main source, the ocean, the atmosphere contains varying quantities of water vapor, (see section I). Water vapor and carbon dioxide are the parts of the atmosphere that store heat.

Variation in atmospheric humidity reflects the amount of water vapor in the atmosphere at any time. For most of human existence, the proportions of nitrogen, oxygen, argon and carbon dioxide have remained constant. However, the percent of carbon dioxide in the air has been shifting. From 1740 through 1840 to 1940 the CO2 content in the atmosphere rose slightly from approximately 278 ppm to 284ppm in 1840 and 310 ppm in 1940. Then it jumped to 370 ppm by 2000.

The "greenhouse effect" is the term used to describe the retention of heat in the Earth's lower atmosphere (troposphere) due to concentrations of certain trace gases and water vapor in the atmosphere. These gases are generally known as greenhouse gases[2]. Concentrations of some of them have increased steadily during the 20th century and into the 21st, with CO_2 rising from under 300 ppm to now approach 400 ppm.

The Making of the Colonial Era:

The early industrial revolution began in Europe's feudal agrarian society. The industrial concerns at that time were related to crop yields and livestock rearing. Crop rotation was introduced to make the farmer's fields more productive. However, in feudal Europe, those, who worked the land were not farmers, but farm laborers. They worked as slaves. Thus, their relationship to the land was negative.

They were the feudal underclass, stripped of a personal sense of human dignity. They worked under the harsh conditions, in which they were often in debt to the respective land-owners. They cultivated the fields, grew the crops, cared for the livestock and harvested the products of the land, all of which generated wealth for the land owners[3].

This was a society in which the privileged and powerful exploited the relative helplessness of the powerless. It was also a society, who saw land merely as a resource, which fed the local population and used its excess wealth to support the local militia, which protected them from the invasion of hostile regional raiding militia.

In general it was a society at war. It had emerged from the chaos of the Middle Ages with a growing sense of colonial expansion, accompanied by unprecedented economic growth through international trade. The resulting competition for global dominance kept them at war with their neighbors and inherently hostile to the land. The period was from 1492 with the Columbus landing to the 1789 breakout of the French Revolution.

The pivotal historical events driving the colonial era were:

- The Doctrine of Discovery and the early European settlement of the Americas,
- The King James Version of the Bible in popular print and the Pilgrim Puritans,
- The Transatlantic Slave Trade,
- The American and French Revolutions,
- The Napoleonic Wars.

This combination of events triggered a wave of expansion from an agitated European population. When Columbus arrived in North America Europe was a collection of countries in a region dominated by the Roman Church. Its leader, the Pope was one of the most powerful figures in Europe. He represented the authority of the Divine Rights of Kings. It was believed that a King, blessed by the Pope possessed unique Divine Rights bestowed on him by God.

The Doctrine of Discovery: Thus, when Pope Nicholas V presented the Doctrine of Discovery to the King of Portugal in 1452, it was for the King a divine directive[4]. So significant was this action that, for centuries thereafter, the principles of Christian mission, as well as civil law were heavily influenced by it. This was the context in which the King James Version of the bible was printed. It was also the first printing of the bible that was available for common distribution.

The Doctrine was also the religious justification for the colonial movement. In effect, the colonial system was a direct extension of European feudal society. It was built and fundamentally supported by an enslaved underclass, who provided the labor necessary to produce and maintain the net economy. As a reward for their loyalty, the indentured servants, who were brought to the Americas, often as crew on the ships that transported a new wave of adventurers seeking new prosperity, were recruited to actually displace the natives. They did their job with brutality to match the way they had been treated as indentured servants in their former lands[5].

Thus, the lands of the Americas had a spiritual transformation from being nurtured by genuine stewardship to being worked by slaves, driven by the brutality of continuous terror. In such a dominant negative relationship to the land, the entire population became oblivious to many of the positive benefits of natural stewardship. The land, the environment and other natural attributes were seen as obstacles to the artificial world order they were being forced to create.

The King James Bible: Today's Christians no longer actively promote the Doctrine of Discovery. The mass printing of the King James Bible exposed lay people and clergy to scriptures in ways that had not been possible before. People could read for themselves what they had been forced to believe second-hand from popes and clergy, who were not familiar with basic and critical concepts laid out in the scriptures. Even now, with all the studies and research available, the force of tradition still influences considerable prejudice and popular misconceptions in the practice of the Christian religion.

The spirit of the Doctrine is so interwoven into much of the structure of North American legal codes that we continue to legally persecute indigenous people. In fact, it remains a living function of the colonial legacy. It is so pervasive that the World Council of Churches filed a complaint with the United Nations in 2012 requesting that the Doctrine of Discovery be officially repudiated[6]. In the current global refugee crisis indigenous people are over-represented

Displaced Indigenous Peoples: In places where the people still lived off the land, their relationship with the land was generally one of harmony in gratitude. They trusted their knowledge and their intimacy with the land and the seasons. They followed the rhythms of the seasons to know when to prepare the land, when to sow the seeds, when to nurture the respective crops and when to bring them to harvest. The appropriate harvest time was often near the end of the summer; a time for harvest festivals and thanksgiving.

The Mayans and Aztecs all celebrated magnificent market festivals. These were popular social and cultural events, often lasting several days at a time, since farmers came from long distances, bearing their best produce. They came to barter with friends, and special acquaintances on significant occasions.

Harvest celebrations were the biggest market event. They were their equivalent of our post-Thanksgiving Day sales in the US. The festivals were generally marked by street fairs, carnivals and a variety of entertaining events. However, the market event itself was what attracted most people to peddle their farm produce, arts and crafts. In addition to regular food items, domestic utensils, farming tools, clothing, baskets for all purposes, crops and all manner of goods imaginable were traded.

The Incas had a unique way of building community. Prior to the arrival of Europeans, (15th and 16th century), the Inca Empire was the largest in South America. They were master city builders, with abundant foodstuffs, textiles, gold and coca. However, they had no open markets, like the Aztecs and Mayans. The Inca government sponsored elaborate feasts to serve as the vehicle, which brought the people together.

The economy of the Incas had been a mystery to historians, until recently. They had food, clothing, household items and other necessities in abundance. They had a different way of distributing and sharing food, clothing and building materials which drove their economic activities[7].

Francisco Pizzaro, the conquistador led just a few men to defeat of the Incan army in Peru in 1532. He was aided by earlier European visitors, who brought along a smallpox epidemic roughly a decade earlier[8].

The trade relationships of the Incas were somewhat personal and intimate, based on trust and respect. Living off the land, as they did, the farmers also maintained an intimate relationship with the land. Inca agricultural yields were significantly higher than anything found in today's modern agriculture[9]. Trade with people outside the Empire was exclusively a governmental operation.

Replacing the Village Concept:

Most traditional societies focused on establishing and maintaining a sense of security, peaceful harmony, cooperative living and a place safe to raise children. The idea behind the statement: "It takes a village to raise a child" embodies this concept. The typical village elders in these societies were deeply committed to their community. Accordingly, elders would deliberate on an issue affecting the village, until they could reach a consensus, even if the deliberation took many days.

The European style of governance involved contesting a wide variety of competing issues. In addition, much of their effort revolved around establishing regulations. The colonists seemed to harbor a continuous array of conflicts, whereas the life in traditional societies seemed to flow more smoothly, according to a collection of stories told on the website: americaslibrary.gov

When one's lifestyle revolves around the land, ongoing stewardship becomes obvious. For modern urbanites, the idea of returning to an agrarian lifestyle may seem like a retreat to the Dark Ages. David Korten of the Living Economies Forum reminds us that the popular pursuit of indefinite economic growth is doomed, because it is not tolerant of universal stewardship and related sustainability[10].

Nevertheless, millions of people of every ethnicity and religion, and every nationality have been applying modern technology in the service of sustainable life options[11]. Some inventions have been around for decades, awaiting the proper support to move them forward. For instance, solar and wind energy have been available and in modest use for centuries. Modern technology is finally making its application more readily accessible to retail consumers. Other ideas, though incomplete, may yet surface with greater recognizable relevance.

Accordingly, for the first time new technologies, such as hydroponics are allowing for more widespread indulgence in Earth stewardship[12]. This innovation will have a positive impact on food production throughout the planet. As we approach the challenges of the climate crisis we can be encouraged by a wide range of new and recent technologies available for the task. We must also be vigilant in our duty to hold the global corporate leaders accountable and reject any attempt they may orchestrate to measures that harm the environment.

Modern communications technology has erased many of the barriers of geography, linking ordinary people into a seamless global web of learning and sharing unfiltered by imperial agents and institutions. The Arab Spring outburst in 2011 and the Occupy movement in the USA that fall bear testimony, from which hope may yet "spring eternal".

In light of these technological developments, David Korten challenges us to "imagine an economy in which life, valued more than money and power, resides with ordinary people who care about one another, their community, and their natural environment. Welcome to the New Economy and the ecology that it brings. Millions of people are living it into being. Our common future hangs in the balance"[13].

CHAPTER 17

❖ ❖ ❖

Basic Stewardship

With, or without the "climate change" crisis, Earth stewardship is critical; not so much for the benefit of Earth, as for our benefit, as humans living on Earth. Earth will continue to be alive without us. This planet is the only place we have access to. It is our **only** home. Earth provides all we need and makes our life possible.

Our more urgent challenge is to get to know and take care of Earth, before we destroy the formula of gasses in the atmosphere that support life as we know it. We have known the composition of the atmosphere, which has supported life on the planet was 21% oxygen, 78% nitrogen, 0.9% argon and 0.031% carbon dioxide in the 1940s.

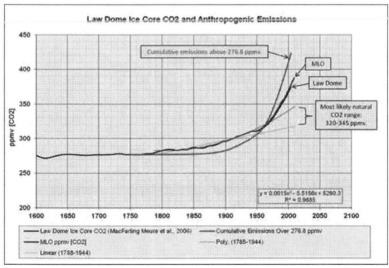

Figure 5. Natural sources probably account for 40-60% of the rise in atmospheric CO_2 since 1750[14].

Today the carbon dioxide share of the atmosphere is over 0.038%. At a glance, this may not seem like much of an increase in 70 years. However, when one takes into consideration that between 1740 and 1940, the increase was from 0.0278%, the more obvious question begins to emerge. What's happening here?!

According to another study, which provided estimates for the period between 1740 and 1940, the 1840 estimate was 0.0284%. The above figure illustrates the dramatic increase in carbon dioxide, (CO_2), in the atmosphere. At this rate, the CO_2 atmospheric content could approach 0.05% by 2100.

Such a CO_2 content has not been known in human history. Thus, we are cultivating a climate, which would make human life, as we know it quite difficult, if not intolerable[15].

The current climate crisis calls for a higher level of international cooperation than what the world had seen before November 2015. For weeks no shortage of speculation swirled around an expectation that this planned global summit on Climate Change would fail, as other efforts had in the past.

The Paris Accord on Climate Change was a Unique Forum:

The first steps toward establishing a unified commitment to decisively reduce carbon emissions were taken in December 2015 in Paris, when 195 nations made a unanimous decision to set goals to reduce carbon emissions in their respective countries. The fact that they achieved consensus is commendable and encouraging. It sets a standard. The standards still have to be followed through. However, a framework has been set to take critical steps toward cutting carbon emissions.

The Consensus Standard: The Paris Accord signaled a clear concept in democratic expression that made the quality of exchange so successful. The current climate crisis has been affecting people in different ways around the world. As such, people had different priorities.

The customary democratic approach of voting for majority rule has only limited success in involving all parties in decisions that favor all concerned. The majority-rule process is primarily confrontational, as it decides winners and losers. In contrast, when we seek consensus, the primary goal is to reach agreement.

It has been generally believed in Eurocentric circles that seeking consensus is too time-consuming. However, the Zulu and Xhosa people of southern Africa have used a forum, called the indaba, designed to enable large groups of people to arrive at consensus. In this process everyone has an opportunity to voice one's opinion. All opinions are heard and considered and the forum eventually produces consensus. This system was used to produce the Paris Accord.

If the purpose of a democracy is to allow all citizens to participate in building community for the common good, then why wouldn't its forum for governance focus on bringing decision-makers to work together for the benefit of all? The Paris Accord forum demonstrated a process of decision-making more appropriate for community governance than the processes currently used by most governments globally. It would benefit the planet, if countries were to apply such a format to govern their respective people.

The Global Corporate Structure Stands in the Way: Modern technology holds the key to the feasible transformation to energy sources that do not add carbon to the atmosphere. The global economy will shift to accommodate such transformation, especially if the corporate giants allow the global market to be more open. The following chapter on Power Generation will lay out the extent to which the global economy may continue, with appropriate responsibility to be profitable without harming the environment.

The carbon-offset strategy that became a popular option a few years ago was not designed to address the urgent need to reduce the carbon buildup as quickly as possible. If one wants to extinguish a fire, one does not merely reduce the fuel being applied to intensify the blaze. The first step would be to cap the fuel input and quickly apply efforts to subdue the blaze[16].

The Global Public Voice: The Paris Accord was a distinct voice of the people of 195 nations to reduce the negative impacts of the current climate crisis. Their effort to articulate the consensus of these diverse nations illustrated to the global community a new approach to political leadership, especially in the context of future generations.

This is how those serious about democratic government would like to be represented. Most neighborhood groups would opt for a system of governance that took their opinions into consideration. They would usually prefer to be able to participate in forums that make decisions for the community. Community gardens tend to function in this representative manner. It is the spirit that aligns with Earth's lifestyle.

The European style of democratic governance seldom achieves consensus, because it is based on a combative model, with the primary goal to declare a winner in any issue under consideration. The consensus approach is focused on getting unanimous decisions. The process, which produced the Paris Accord demonstrated such feasibility. It is now up to us in community to demand such leadership and governance[17].

For those, who cannot imagine consensus in government as valid and for those, who do not think certain citizens deserve the opportunity to participate in their restrictive electoral processes, the Paris Accord was merely the wishful thinking of misdirected environmentalists. Their comments were swift and to the point[18].

Nevertheless, there are many corporate leaders, who have a vision for future generations. Fortunately, these are the long-term corporations, which have invested in future generations. We must support and encourage them[19].

Entire established power structures are heavily invested in the present popular forums, currently in place. However, we already use the more democratic consensus forums in many familiar institutions mentioned above. We must continue to indulge in these community-building forums in our more neighborly activities.

Given the current global political leadership, the only way to ensure the participation of the United States and China in the development of the Paris agreement was to make it nonbinding. The Obama Administration insisted on it, well aware that the U.S. Senate wouldn't ratify a formal treaty. China, which has long insisted that countries should be allowed to tackle climate change in their own ways, sided with Washington.

Nevertheless, future generations deserve far more than what we have prepared for them at this point in history. Our grandchildren need our participation and practice in using such forums in community governance, among our non-profit agencies, our churches and civic organizations, so that, with practice we would become more comfortable with the use of consensus decision-making.

The Paris Accord sets the task of global stewardship in motion. We all need to participate in the processes ahead. Accordingly, stewardship must be pursued with appropriate urgency.

Community Building is Necessary for Effective Stewardship:

Civilization implies a predominance of civility within society. Yet, the post-colonial modern lifestyle has inherited an artificial segregation, in which the European population has usually separated itself from the non-Europeans in the former colonies. With it, the post-colonial European examples of democracy, which is essentially combative, has also thrived. It has evolved into a system of perpetual lock-horn struggle among leaders, who have ignored the public in their contrived battles ignoring the voice of their constituents.

Community building through these new inner-city community forums seems to, (both formally and informally), adopt cooperative leadership in their processes. It is important to them that everyone feels a sense of belonging, based on one's sense of self-value in the respective neighborhood activities.

The village culture of the community garden often has a rural flavor, with most people in tune with Earth's lifestyle. In the community garden setting, for instance, the raw urban instinct to compete for everything is replaced by the more obvious need to work together on such routine tasks as watering the field, helping each other with harvesting chores. Children grow up seeing neighbors helping each other and sharing crops.

Out of this type of community-building in neighborhoods, the temptation to steal and rob becomes less attractive. In fact, community farming groups sometimes cooperate to stage joint farmers markets. They sometimes use block parties to enhance their presence in their respective neighborhoods.

In today's urban society, cooperative societies duplicate such a community expression within their urban settlements. Cooperative housing often provides the village-style interaction, which provide mutual strength to cooperating groups. As we will discuss at the end of Section IV, urban community gardens have incorporated many of the essential community building customs to return a sense of community among the poorest of neighborhoods[20].

CHAPTER 18

Power Generation

TODAY'S GLOBAL ECONOMY has been built around energy. Modern technology depends increasingly on the availability of a reliable energy source. The most important industrial systems are, at least, partially computerized and large computers require an uninterrupted electric power source. The most reliable continuous electricity source for urban industrial consumption depends on a reliable electric power grid.

The electric grid in the USA is powered by a variety of natural resources. These include, (but are not limited to): Petroleum, coal, sun, wind, running water, geysers and other volcanic sources. The electric grid is the most critical unit of infrastructure in the matrix of systems upon which we depend in our modern habitat. Electricity also powers many important parts of overall infrastructure, such as: water, sewer, roads, bridges, dams, levees, etc.

Future Industry will be Driven by New Power:

In post-World War II the USA experienced an economic boom unlike anything in prior human experience, (see Chapter 21 – The Crumbling Infrastructure). However, future demands in a world adjusting to renewable energy, the switch from coal, oil and natural gas will require more efficient operations to be smaller and more locally powered.

In this new post-petroleum efficient corporate operations would revolve around concentrated solar power plants, or wind turbines. A few of the European countries have already made great strides to liberate themselves of petroleum tyranny. Most new countries are without economically viable stores of coal and petroleum. Those, who have will increasingly find it necessary to develop their solar capacity for future energy security as the current technology to support wider use of the always available solar options. Current global energy use still looks much as it has over the past three or four decades.

Carbon-Based Energy Fuels:

Carbon-based fuels are the most common sources of electric power used in today's industrial world. Petroleum and coal have provided most of energy used in the past century. However, in recent years natural gas has risen to prominence globally. Compared to coal and petroleum, natural gas is a cleaner fuel. However, extracting it has incurred major disturbance to land stability, as well as seriously contaminating local ground water in the process. This will be discussed later in appropriate detail.

Oil and coal have been the prime contributors to the green-house conditions most responsible for the rapid increase in global warming over the recent decades. Those industries have, in addition, contributed significantly to widespread pollution and contamination of waterways and oceans. The extraction and transportation of both products creates significant pollution on land and crude oil at sea.

Employee safety is significantly ignored in the industries related to energy production. Accordingly, the most disastrous industry "accidents" have occurred, (and continue to occur), in these industries.

Corporate Manipulation of Environmental Crisis Information:

In 2003, in the face of rapidly accelerating global warming, President G. W. Bush appointed a group of pseudo-scientists, who created a "think tank", called the "Competitive Enterprises Institute". Their primary objective was to confuse the public by claiming that global warming was merely an empty theory and a hoax. These pretenders were funded by corporations from the oil and coal industries to discredit the science of climate change. The additional goal was to counteract the need for public alarm over the impact of energy over-consumption on the environment[21].

Encouraged by the President's chosen spokesmen, and strict censorship of the administration between 2002 and 2007, the public was easily confused by the false claims courted by the government on behalf of the industries. Since then, reputable scientists are finding more compelling evidence through continued research on the present and future potential impacts on the planet[22].

Nowhere else in the industrial world would this have these pretenders been taken seriously. The value of the widespread ignorance of the US public made this confusion possible.

If the coal and petroleum industries had stakes in confusing the public on the environmental issues related to climate change, in the aftermath of the Presidential election of the USA in 2008 they kicked off a new concerted effort to keep the funding sources secret[23]. Below is a graph representing the way their funding sources have been hidden from routine public access.

The temporary advantage of surprising the public with a sense that global warming may not be as serious as previously indicated has served to comfort much of the public that there is no need to be concerned about our modern lifestyle. However, truth has a way of rising to the surface of any controversy[24]. Soon, the world could identify most of the money supporting these bogus claims was being provided by the coal and oil industries. They had to go underground with their funding sources.

The Coal Mining Hazard:

Underground mining has been hazardous to miners in many ways, for centuries. Since the industrial revolution was born in Europe, such hazards have been well known. However, the USA has had ample experience during the past 150 years to verify the acute danger of such built in occupational hazards. The United States Mine Rescue Association has kept extensive records and a running account of disasters, which reveal a grim picture.

Climate denial funds from fossil sources

Amount given to climate denial groups

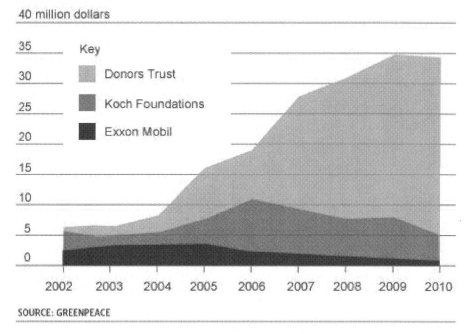

SOURCE: GREENPEACE

Graphic: climate denial funding The Guardian, Thursday 14 February 2013

Given this past record, it may surprise reasonable readers that events, such as the mine disaster of April 2010 in West Virginia had been so quickly forgotten. The following statistics, taken from the United States Bureau of Mines may offer hints at a trend in media coverage that seems to persist.

Table 1: Statistical Summary Number of Documented Mine Disasters, (where 5 or more deaths occurred):

Historical Period	Coal Mine Disasters	Metal & Mine Disasters	Total Disasters
Thru 1875	19	4	23
1876-1900	101	17	118
1901-1925	305	51	356
1926-1950	147	23	170
1951-1975	35	9	44
1976 – present	16	1	17

Year with largest number of coal mine disasters: 1909 (20 coal mine disasters)

The above table identifies the number of mine disasters occurring within the respective periods. It is clear that the frequency in disasters has decreased since World War II. During the three years leading up to the establishment of the U. S. Bureau of Mines, 1907 to 1909, there were 50 coal mine disasters in which 5 or more miners were killed. The total killed in these mines was 1,773. That works out to an average of just under 35 per mine.

Table 2: The Five Worst Coal Mine Disasters in the USA:

Date	Mine	Location	Disaster-type	Deaths
6 December 1907:	Monongah, #s 6 & 8	Monongah, WV	Explosion	362
22 October 1913:	Stag Canon, #2	Dawson, NM	Explosion	263
13 November 1909:	St. Paul, #2	Cherry, IL	Fire	259
19 December 1907:	Darr mine	Jacobs Creek, PA	Explosion	239
1 May 1900:	Winter Quarters, #s 1 & 2	Scofield, UT	Explosion	200

The five worst coal mine disasters in the USA since 1940 were:

Date	Mine	Location	Disaster-type	Deaths
20 Nov. 1968	Consol,#9	Farmington, WV	Explosion	78
21Dec. 1951	Orient,	West Frankfort, IL	Explosion	119
25Mar. 1947	Centralia #5	Centralia, IL	Explosion	111
16 Mar. 1940	Willow Grove #10 St.	Clairsville, OH	Explosion	72
10 Jan. 1940	Pond Creek #1	Bartley, WV	Explosion	91

Coal Mining Disasters.

The lives of miners are taken for granted. Over the past two centuries, (and indeed throughout the industrial revolution), the great risk faced by miner is legendary. Like the whalers of previous generations, who were sent out to provide their customers with fuel oil for their lamps and the many other uses, they have provided for us at great cost, risking their lives. These are the quasi-slaves who have built our urban industrial lifestyle.

Over space and time, our sense of "independence" is shown to be false. We are completely dependent on others, who often risk their lives to manufacture many of the basic goods we buy and take for granted. We also owe a huge debt of gratitude to the lives of countless underpaid and over-worked people, who sacrificed their lives to make our goods inexpensive. In this way, their poverty makes much of the urban experiment possible. Their sacrifices make goods "affordable" to us. Our real life is one of continuous *INTERDEPENDENCE!*

Mining of all types carries grave risks. One has only to recap the drama of the rescue of the 33 copper miners in Chile in October 2010, to appreciate our dependence on such horrendous sacrifice on the part of many unseen victims. They had somehow survived in a cavity 700 meters beneath the surface. According to Jonathan Franklin of the Guardian, in his report from Santiago, Chile on August 4, 2011, on the anniversary of the mine collapse, many of the rescued miners were living in poverty and facing a grim future[25].

This tragedy echoes an unfortunate trend in mine management. The mine operators often fail to correct the potential hazards to their workers' lives. Thus, the working conditions are both hazardous and uncertain as people are forced to work, amid the fear of entrapment and possible death. Working conditions are usually strenuous and uncomfortable.

In addition, they had to survive the constant heat that results from being deep in the underground cavities. Under normal conditions, the temperature increases as one descends into the earth. Also, the underground is subject to continuous water seepage from the natural ground water that constitutes the water table everywhere, (even in the desert, if one goes far enough below ground). Such conditions exist wherever mines are dug.

Apart from these natural conditions, abuse of miners often grows out of the negligence of management. Follow-up reports from the Chilean mine disaster indicated that parts of the emergency ladder, which every mine in Chile is supposed to have, had not been completed. In fact, the miners had climbed one-third of the way up to that point, where they remained stuck for 33 days[26]. The above reflect the hazardous life of miners the world over.

Other Coal Mining Hazards: In addition to the hazards of working in the mines, the mining process presents other challenges to the physical environs around the mines and to power plants that use coal.

Depending on the type of mining methods practiced, considerable earth material is extracted on order to release the useful coal. The resulting waste material, (slag), in storage has been a source of considerable inconvenience to mining companies. In addition, it presents a major hazard to the surrounding community in many ways. Where the mine has been in operation for many generations, slag heaps can become a monumental problem!

I studied Economic Geography at the University of Durham, England in 1966-1967, when coal mining was still a major industry in the area. That local coal mining there goes back two centuries.

During my stay there, a disaster struck in the village of Aberfan in Wales, where for, 50 years, the existing mine operation had built a mountain of slag material. It had rained for an entire week previous, in the area, soaking into a slag heap on a hillside, directly above the village, in the valley below. The mountain of slag, several hundred feet tall collapsed, creating a massive avalanche, which wiped out a school and several houses, killing 116 children and 28 adults. Warnings of the potential disaster had been voiced and recorded for years, but not heeded[27].

It was many years later, when I discovered that a friend, whom I had known only as a successful engineer, with his own business, told me of his early life as the son of a coalminer. Hearing him talk of his father coming home so exhausted, covered in coal dust that his mother had to clean him up in a tub before he could sit down to eat and even then he often would fall asleep while eating!

My friend, almost in tears himself, would say with some lingering distress that his brothers saw him as something of a traitor to leave home and never return to take his expected place as a coalminer's son! On the other hand, this tragic past was his incentive to study extensively and get scholarships to pursue an excellent education, so that he would never have to live that type of life.

The Epidemic of Work-Related Injuries in the USA:

The Center for Public Integrity has recently completed an in-depth study on the chronic nature of official neglect of workers exposed to routine occupational hazards on the job. Of course, these are common manual laborers, (including miners), whose jobs have traditionally been unprotected throughout most of the duration of the modern industrial era. The study also observes the impact that this level of non-protection has had on the families of injured or killed workers[28].

Mountaintop removal:

Mountaintop removal is a quick and easy way to mine the seams of coal found in the Appalachian mountains of Tennessee and West Virginia. The companies blast off the tops of mountains to get at the seams, using the open-pit method to excavate in a few days what it would take weeks of manual labor weeks to excavate, had it been done un-derground by earlier traditional means. The cost of that convenience is exceedingly high in lives, worker casualties, water contamination, de-struction of property, avalanches, flash floods, public health, etc. Such costs have been borne mostly by local residents, who live in the valleys, into which these mountain streams drain.

Consider these predictable consequences of imposing mountain-top removal on the geography of drainage and erosion:

- Each mountain is drained by several rivers and streams.
- When it rains, (or snows), on a mountain, water, (including snow-melt water), drains downward, following the natural valleys toward lower ground.
- Where water flows, it tends to erode loose material and transport it downstream.
- Excessive rainfall in an area of significant loose material is likely to produce flood conditions, in which loose soil becomes saturated and flows as thick mud along the streambed.
- Such mud-flows are sufficiently strong to destroy buildings and/or transport homes downstream.
- When a mountaintop has been broken down by explosions much of the debris is fine, powdery material, which easily becomes suspended in the flow of a fast-flowing stream.
- Where the rock base of the mountain contains large quantities of sandstone and/or limestone, the fine-grained material can dissolve easily in water to form a mudflow almost the consistency of wet cement, so that, wherever it is allowed to stand it can solidify.

Villages, small towns and other settlements along the valleys downstream from the removed mountaintops have been inundated by flash-floods and other water borne disasters seldom seen before. According to Appalachian Voice, in a report on 29 September 2006, over 500 mountain tops had been removed. Some of these explosions remove as much as 400-800 vertical feet from the tops of mountains. Many such explosive removals result in the damming up of streams. Other situations leave entire valleys filled in. EPA studies have sited water pollution as a major health hazard. Unsafe quantities of arsenic and other life-threatening pollutants have been identified[29].

Many local law-suits have been filed against delinquent mines and they move slowly through the court systems. In the meanwhile, very few disasters are punished and few affected miners and their families are appropriately compensated, while the environmental and health abuses continue[30].

Coal Ash Dams:

Coal powered electric power plants present toxic threats to local communities that rival oil spills in potency. The common practice is to store the coal ash generated when coal is burned for power. The remaining ash may be in liquid form, or solid form. The liquid form is collected in large surface impoundments or dams. The solid ash goes to landfills. Both deposits are supposed to be protected, because ash contains contaminants like mercury, cadmium and arsenic, which can cause cancer and other serious health problems[31].

The Institute for Southern Studies, in Durham, NC has been monitoring coal ash dams. There were 137 locations across the U.S. where environmental contamination from coal ash disposal has been documented by EPA, when they announced that some 1.5 million children were living in contaminated areas in March 2011. On 17 May 2011 a final EPA release identified 676 sites that had been inspected, of which, 55 got a "poor" rating, meaning they need repair or further testing, or lack proper engineering documentation[32].

The dams are of particular concern, because dam breaks are too common and their failure inflicts considerable property damage on poor communities. Furthermore, the health and environmental hazards that accompany these events are often worse than most "natural disasters".

So many of these coalmining disasters are kept in low profile that the public continue to miss their collective impact. In contrast, the image of "clean coal" seems to make a bigger impression. The above contaminating events are left behind in the process of cleaning the coal for our "clean coal" market[33].

Big Sandy River. (Photo © Wade Payne courtesy Greenpeace)

Other Dam Experiences:

When digging underground, sooner or later one confronts the water table. Because owners and managers do not have to venture into this working environment, protection of the workers who are assigned to such working conditions is generally minimal. Such working environments expose and tend to concentrate hazardous materials not generally found in our living environment above ground. When the desired minerals have been completely extracted, mines are usually abandoned.

When mines are dug into the side of a mountain, draining the excess water can be easily drained out using gravity. In such cases, the drainage naturally enters streams. This is the source of easy contamination of such streams. Regulatory agencies are likely to monitor the drainage from the mine, if it is known to significantly contaminate streams being used by a significant population downstream. However, the staffing of dam monitors and inspectors across the USA is woefully inadequate, (as is discussed in more detail in Chapter 22 on Infrastructure).

The dam collapse freed tons of iron ore and silica mud contaminated with heavy metals such as mercury and chromium © Ricardo Moraes/Reuters/ Corbis. [Compare it with the level of the normal flow of this river and appreciate the scale of devastation represented here. Observe the mark left from the flash flood created by the dam break several miles upstream.]

The environmental disaster that followed the collapse of a dam at a Brazilian mine on 5 November 2015 has caused unprecedented damage in that country and will have irreversible negative effects on human health and the environment, according to experts[34].

At least 11 people have died and more than 600 were displaced. In addition, the water supply of more than 250,000 people in the area was contaminated with heavy metals, such as: mercury, arsenic, chromium and manganese at levels exceeding drinking water limits. Thus, water had to be cut off to the residents. Tons of mud made up of iron ore waste and silica, originally estimated to be about 25,000 Olympic swimming pools in volume, have spread over 800km and devastated one of the largest Brazilian rivers, the Rio Doce. The contaminated mud has reached the Atlantic coast and could potentially impact the wider marine ecosystem, (see Section II).

Art Levine, in a report to Truthout on Monday 12 April 2010 described the risky situation left by the Leadville Mine, Colorado. At the time a tunnel had collapsed in an abandoned mine, trapping a lake behind it. The Leadville Mine Drainage Tunnel, (LMDT), had been built to drain the mine. But, since the tunnel collapse, a contaminated pool behind the collapsed debris was building up. This left the Bureau of Reclamation (BOR) and Environment Protection Agency (EPA) with a disconcerting problem.

Mr. Levine commented:

"We are often confronted with the realization that what we thought we understood remains a mystery to our scientific "know how". What do we do about the fact that the water keeps coming, filled with the contaminants we released, when we blasted stable sediment to release the mineral source we wanted, not knowing what impact it would have on the other needs of the entire community. But, we quietly move on, as if nothing happened, leaving the mystery to surprise some other scientist later, who did not know that a red flag had gone up years ago, that had not been recorded, or assessed to pass on the extent of what we collectively did know.

Thus, in our own scientific world, the negligence of a few affects the decision making of future generations, compounding the errors we would generate in the future, preserving more misunderstandings of the flat earth[35].

After months of planning to avert a perceived disaster at the LMDT due to high water levels backed up behind a collapse in the tunnel, the BOR has drilled a hole into the tunnel, placed a pump inside it, which now transports more than 1,000 gallons per minute from the tunnel to a treatment plant before releasing the water into the Arkansas River.

The Petroleum Industry:

Today's energy crisis is not a shortage of petroleum, but the fact that we consume far too much petroleum, too quickly. Furthermore, we waste too much of what we consume.

The USA consumed almost ¼ of the current global oil. Many industrial European countries, Australia, Japan have populations, which demonstrate much more cooperative, harmonious lifestyles. They all provide a larger percentage of their respective populations with adequate food, clothing, health, education and a sense of general wellbeing than what is being experienced in the USA. And they do this using far less energy. In fact, with 5% of the global population, the USA consumes 20% of the global energy consumption.

More than half the energy used in this industrial era was consumed in the last two decades. Much of that increase has been in vehicular traffic. Forty percent of the oil used in USA is used to power personal vehicles, 70% if you include trucks. Thus it seems that in the last two decades, we have done more damage to our environment than in all previous human history!

Available solar energy far exceeds all the other available energy resources on Earth. This should be good news to us. However, corporate greed stands in the way of seriously pursuing this option in the interest of our grandchildren. Current technology does not allow us to harness transferable energy at the rate we currently use it. In contrast, solar energy does not require large central power plants. It does not require transportation and is far more efficient than any energy available from today's power plants.

Oil extraction, by its very nature, usually requires a simpler process of retrieval than coal. Petroleum is usually pumped up from underground as a liquid. It can be transported in pipelines, (as crude oil), directly to refineries, or into vessels to be transported, (usually,) overseas, to refineries that turn it into petroleum for further distribution, wherever needed. However, transportation carries major risks:

Super-tankers have multiplied the risk in recent years.

Under-water drilling for oil has presented significant problems.

Deep water drilling intensifies the risk and the technology is not fully developed for safe execution of such projects.

Oil spills on the open ocean are often disastrous.

To date, oil companies responsible for the respective spills have not often been prompt, or responsible about cleaning up, especially when the spill is not close to dry land.

Natural Gas and the Fracking Dilemma:

Shortly after a gas company in Donegal, Pennsylvania, began storing fracking wastewater in an impoundment pit, a water well at a nearby home showed some alarmingly elevated levels of barium and strontium. Other chemical pollutants, included arsenic, chloride and radium[36].

In addition to the chemicals released into ground water during hydraulic fracturing, (fracking), large quantities of water are required. Water plays a major role in the shale gas recovery process, as well. In order to release the gas from shale rocks large amounts of water, fine sand and chemical substances must be injected under high pressure in the ground. Hydraulic fracturing requires as much as five million gallons of water per well according to the International Energy Agency (IEA). In some areas of Texas, communities have seen their groundwater supply virtually disappear overnight, following a fracking episode[37].

Then came the earthquakes: In recent years, indications of swarms of earthquakes began to surface in the regions of large volumes of fracking events. At first, these quakes averaged around magnitude 3.0 on the Richter scale. By 2009, it became clear that they were being caused by the fracking. However, the oil industry managed to refute these claims. However, when a magnitude 5.7 was recorded, that was too ominous to be ignored. Katie Keranan, an assistant professor of geophysics at the University of Oklahoma, partnered with scientists from the USGS and Columbia University's Lamont-Doherty Earth Observatory to install two dozen seismometers in Prague, Oklahoma[38]. These developments have prompted more direct concern about future fracking activities. In the process, a more concerted opposition to future fracking has arisen among state and local environmental advocate groups.

Non-Carbon Based Energy Fuel:

Throughout human history, energy has been obtained from a wide variety of sources prior to the discovery of coal and, later petroleum. In addition, hydro-electric power, nuclear energy, tidal energy, many forms of thermal energy are currently in use. Much energy is generated by other recycling applications. Many operations, which create large quantities of steam, super-heated steam, hot water are devising ways to reuse the "waste energy". Some garbage collectors are designed to separate out organic waste, hold it in a confined space and recycle the heat generated by the composting process.

Yet, the most widespread and easily available source of energy continues to be the sun. It also has been in use longer than any other source. Solar sources remain the most available and sustainable.

The greatest value of solar energy to the consumer is that it is available in locations that are off the electric grid. Passive solar energy is available without equipment, although such applications are limited. With little more than a reflector and a box, one can bake bread and perform other types of cooking tasks. The reflector application has been used recently in many rural villages that are off the grid in India. In fact, in many such villages, off the electric grid, the use of the computer has flourished, nonetheless.

The wind is being used to harness power in many strategic locations. In places on the ocean front, where tide changes are significant, the tide is being used as a reliable source of energy. Rivers can sometimes produce hydroelectric power. In many areas, where significant volcanic action beneath the surface, geysers develop and produce sufficient thermal pressure to support a thermal power plant.

When a pipeline breaks in the forest,
and nobody is around to hear it,
does the oil spill
and pollute the groundwater?

From: The Piece Fit, www.piecefit.com

Geyser-harnessed Power:

Wherever geysers are found, there is a magma chamber at some depth beneath the geyser. The phenomenon is made possible by ground water in the soil. Over time, the movement of water on the ground above the magma chamber has absorbed the chemicals present in rock sediments. The heat from the magma chamber heats the mixture of water and chemicals collected in the enclosed cavity below an opening to the surface.

The pressure allows the ground water to be heated so rapidly that it expands, begins to turn to steam, which gushes out of the narrow opening as a mixture of steam and super-heated water. More of the water turns to steam, as it comes into contact with the atmosphere. The steam from the discharge condenses back into water, which drains back into the cavity from which it came to begin the heating process once more[39].

Sometimes, as the pressure is released in the chamber, a residue of chemicals congeals and solidifies to temporarily plug the vent at the surface once more. Thus, the stage is set for the eruption to repeat itself. In Iceland, the country's electricity is generated from such geyser activity. This resource is renewed almost instantly. It is, therefore, eminently sustainable as a useful resource.

The Iceland experiment:

Iceland is located on the Northern part of the Mid-Atlantic Ridge, which is a continuous ridge of volcanic mountains under the surface of the Atlantic Ocean. The ridge runs from the Arctic region of the ocean to the Antarctic. Iceland represents the only part of the ridge above sea level. The Mid-Atlantic Ridge is one of the most geologically active place on Earth.

The country, Iceland, takes full advantage of its geography and geology. They get 81% of their electricity from their natural, renewable resources. Of this locally generated electricity, 30% comes from geothermal and 70% is derived from its many glacier-fed rivers and waterfalls as hydro-electric power.

In the 1940s, the government launched the State Electricity Authority in order to maximize the utilization and management of geothermal energy. With steady growth, it has become a very successful and economically viable development. Today, the national government and the City of Reykjavik jointly own the Authority, which has become the National Power Company and has taken control of the country's geothermal industries, complete with cutting-edge technology and tested application[40].

The development of this service and related industries has met significant challenges. Iceland gets its geothermal energy from some 200 volcanoes, 600 hot springs and 20 high temperature steam fields. Volcanic eruptions are an ever present reality in Iceland. Most eruptions are mild. However, a major eruption in April 2010 grounded air transportation throughout most of Europe for weeks. Volcanic ash is made up of very fine particles, which can get into Jet airline engines and cause airplanes to crash.

The most immediate problem to local residents in the vicinity of a volcano is the ash that can be a hazard for people sensitive to air pollution. Thus, people are sometimes limited in their mobility when ash is in the air. The 2010 eruption spewed a huge volume of ash which spread out over much of the island east of the volcanoes, covering houses, roads, cars, open fields. Such inconveniences could be costly and persistent.

The USA generates the most geothermal energy among countries worldwide. 3,086 megawatts are produced from 77 power plants. The greatest concentration is found at The Geysers in Northern California.

In 2000, the world's production of geothermal energy was approximately 16,211 megawatts, (MW). The USA generated 5,336 MW, or 32%. In comparison, China produced 2,814 MW, (17%) and Iceland, 1,469, or 9%. Turkey moved into fourth place that year, when their extensive program went on line[41].

Nuclear Energy:

When we first discover the enormous power of nuclear energy, we were certain that it was our ideal energy resource. It prompted one of those unbelievable promises, which we expected to take us into space and beyond. Nuclear energy, however, is like a tsunami, the promise of whose power we crave. However, our best science has never been able to handle the scope and volume of power that is actually produces in a nuclear reaction[42].

To date, this source of energy has been provided in many parts of the world. Nevertheless, it comes with, by far, the most expensive start-up costs for plant construction and initial operation of any energy generating venture. Its preparation has been so hazardous, that specialized commissions have been appointed to oversee their progress from the initial application to eventual full construction.

However, the experiment was never completed. Attracted by the power produced by the experiment, the industry took off, leaving incomplete the side of the equation that is supposed to close the experiment loop safely. Thus, the cocktail party was launched on deck to praise and celebrate some of the researchers, abandoning the other researchers in the hold below trying to complete the loop, returning the atom to its original state. The mess we left behind is called "nuclear waste".

The greatest risk in ongoing operations comes from the still incomplete experiment. It is like managing a city without a recycling facilities for its garbage collection system and without the space to safely dispose of the unrecycled garbage. The risk lies in nature of hazardous nuclear waste. The long-term risk is that the half-life of radioactive materials causes it to be extremely hazardous for 250,000 years.

"No problem!" said the nuclear industrialists at the height of the party on deck. They obviously had not heard about the 250,000 year half-life. "Yes! I heard about it, but I won't be around then, neither would you! Let's just store it in a thick vault somewhere out of the way!

However, a part of that risk has been leakage of hazardous nuclear waste during transportation of such waste from operating units to designated waste storage locations. And there have been many leaks.

From an environmental view point, the operation of Nuclear Power plants is not yet safe for power generation. It is the only energy-generating process without sufficient technological knowledge of a safe disposal method for the hazardous waste left behind[43].

Three Mile Island, Chernobyl and Fukushima stand as scary reminders of the tangible risks inherent in the industry. Why the inherent risk? Because, according to findings in the research and routine monitoring of the Nuclear Regulatory Commission, (NRC), many near-miss incidents in plants across the country indicate times, when we have been close to similar disasters[44].

The most up-to-date underground repository was built near Carlsbad, New Mexico in 1999. It is a 98-foot-wide, two-mile-long ditch with steep walls 33 feet deep that bristles with magnets and radar reflectors will stand for millennia as a warning to future humans not to trifle with what is hidden inside the Waste Isolation Pilot Project (WIPP). It is the nation's only deep geological repository for nuclear waste. The site is 2,150 feet underground, carved into a salt bed, 2000 feet thick, formed 250 million years ago[45].

The human cost of operating such a facility is not worth the real and potential hazards generated in the usual plant operation. Workers in nuclear reactors and similar facilities are exposed to unusually hazardous conditions. In fact, reactors operate in a working environment so hazardous that, since the Three Mile accident of 28 March 1979, no new plants have been started in the USA as of 2012.

In 2011 the Associated Press, revealed, (based on their own investigation), that the NRD had been "working closely with the nuclear power industry to keep the nation's aging reactors operating within safety standards by repeatedly weakening those standards, or simply failing to enforce them"

The prevailing argument/rationale goes like this: In the wake of the Fukushima disaster, authorities would down-play any warning to avoid creating a public panic[46].

Nuclear plants are not only most expensive to start up; in the USA experience 10 % of operation time is lost in "unplanned shutdowns", due to miscellaneous irregularities. Nationwide, some 75% of the nuclear reactors "routinely leak radioactive tritium". Where such leaks are a significant public safety threat, a plant shutdown should occur. Each shutdown is expensive and cumbersome.

In addition, during the last 60 years, over 50 nuclear disasters have occurred worldwide. These disasters extend severe risk to employees. At least 11 workers have been killed in US reactor accidents[47].

The nuclear energy industry has been losing millions, because of the exorbitant cost of running a plant with reasonable safety to its employees. In addition to plant safety considerations, we, the community need to keep abreast of industrial activities in our neighborhoods. For instance, Diablo Canyon, a two-unit nuclear power plant in California is located near four known major earthquake faults and just 45 miles away from the San Andreas Fault.

This location has been a major concern of activists opposed to its construction, since 1968, when construction permits were issued. However, the impact that such information has on people with different stakes in the projects in question vary. Indeed, those in opposition may become increasingly indignant, while those intent on continuing may become more determined to push ahead. The resulting dynamic always serves to further delay action on the dilemma.

These concerns were increased in 2008 with the discovery of a previously unidentified fault line that runs about 300 meters from the reactors' safety-significant cooling water intake structure. The new seismic data projected earthquake scenarios for the Shoreline Fault and later the San Luis Bay Fault and the Los Osos Fault that are each capable of producing ground shaking forces greater than what Diablo Canyon safe shutdown systems are designed for[48]. As of 2014, virtually nothing has been done in response to the obvious danger posed by these earthquake faults.

Our Deficient Collective Memory:

Why are these details important? We have a notorious community collective memory, (see Chapter 15 on floods). For example, who remembers "The Donora Smog" of 1948? We can learn much from that disaster and the extent to which disasters are facilitated by the disregard for workers' safety and public welfare by corporate leaders.

The Donora Smog: Some time, between Oct. 26 and 31, 1948, 20 people were asphyxiated and over 7,000 were hospitalized or became ill as the result of severe air pollution over Donora, Washington County, a Monongahela River town of 14,000. During those several days, a critical temperature inversion had trapped the air in the valley beneath a dense cloud cover. The Donora Zinc Works and the American Steel & Wire Co. plant, both operated by U.S. Steel Corp. and situated side-by-side along the Monongahela River bank, had been emitting airborne pollutants from the zinc smelting plant and the steel mills close to the ground. The weather conditions in the valley began to affect many residents with respiratory ailments. As the condition persisted, the source of the pollution became clear. Yet, it took five days, eleven deaths and thousands to be hospitalized before the plant operators could be convinced to shut down plant operations. By the end of this ordeal nine more deaths had been added to the initial victims[49].

The Donora disaster yielded one significant impact on national air pollution standards. It gave rise to local, regional, state and national laws to reduce and control factory smoke and culminated with the nation's Clean Air Act of 1970. However, for people living under such conditions, they can get used to breathing polluted air and tend to lose the capacity to recognize when conditions become unsafe.

Yet, it is not the fault of the people in hazardous areas. When the government allows certain conditions to reach such a level of disaster, without forcing a killing plant to be shut down immediately, the authorities are squarely at fault for allowing such operations to seem safe and acceptable.

Almost 31 years after the Donora Smog event not far away from that site, we had the Three-Mile disaster. That event made it abundantly clear that the Atomic Energy Commission's regulations were inadequate.

Justin Shawley, the high school student researched the 1948 incident and promoted the placement of a commemorative plaque as a marker to dedicate the memory of the victims of the tragic air pollution incident in the Washington County community of Donora. The plaque was laid on Oct. 28, 1998 by the Pennsylvania Historical and Museum Commission to acknowledge the 50th anniversary of the tragedy[50].

Global regulations have been improved, as the nuclear industry continued to learn on the job. The Chernobyl incident of 1986 reminded us, however, that the Nuclear Industry was still unsafe. As such, avoiding public panic is not a legitimate excuse to misinform the public of potential public disaster.

Moreover, it is a troubling sign that the news media, (the folks who claim to risk everything to give us all the news we have a right to know), conspire to keep the public in the dark about such critical situations. Such attitudes are a foolproof formula for building public mistrust.

The Fukushima disaster a few years ago signaled the loudest warning so far that nuclear energy, as we know it today is unsafe and prone to produce large-scale disasters. Incidentally, it is important to know that General Electric designed and built the Fukushima plant in the same design as 23 such plants built in the USA. Three such plants were shut down by the Superstorm Sandy in November 2012 in the New York – New Jersey area[51].

The surrounding community near the reactor can quickly become victim to reactor accidents. Radioactive fallout from the Chernobyl meltdown reached Sweden and other countries. No other energy source in use today carries such international consequences. The hazardous waste left behind by the operation of nuclear reactors cannot be stored safely into the distant future[52]. Earthquakes, volcanoes, earth movement, major erosion and other geologic disturbances can, in time, expose anything we seek to bury underground. Nuclear waste poses a potential disaster for over 250,000 years, the scale of which we have no experience. The rock cycle is not a scheduled event! In 250,000 years a lot can happen. Furthermore, who are we to plan for such a future?!

Locating nuclear power plants in developing countries is even more disturbing. These countries cannot afford the exorbitant cost of adequate ongoing oversight of the plant's operations. Furthermore, the unscrupulous handling of safety details demonstrated by available international experts has proved unsatisfactory in the case of the Fukushima disaster. The record of safety standards by the industrial countries in the developing world should ban them from any future activity outside their own country into the foreseeable future, when the developing countries can be adequately protected. Current systems do not accomplish such.

The Solar Energy Gift:

The major crisis in the use and application of solar energy is the extent to which it continues to be ignored as a primary energy source. In our pursuit of critical energy reserves, the generation of solar energy lags far behind the scarce and expensive oil and gas exploitation. By far the most expensive and hazardous energy source is the harnessing of nuclear energy.

The sun is available to everyone, free of charge. Compared to coal, natural gas, petroleum and nuclear energy sources, the generation and use of solar energy is relatively simple. Its generation is also significantly less hazardous to both workers and those who live and/or work near to the respective energy generators. Furthermore, its greatest fundamental value is that solar energy is the most available energy resource globally . . . . . and it is free!

Anyone can harness solar energy, where the sun shines. Simple collectors can be used for cooking, baking, heating water. At relatively low cost, photo-voltaic cells can be used to harness energy from daylight, even in the absence of direct sunlight.

This does not mean that hazards do not exist in the generation of solar energy. Wherever electricity is being generated, technicians have to take the customary precautions with wire connections and other routine handling of electrified components. However, the relative simplicity of installing solar systems on site for small operations allows systems to produce power within a day. The same is true for wind-generated electricity.

Wind power is generated from solar energy, in that, winds are produced by the unequal heating of the planet's surface by the sun in different locations. The primary hazard related to wind installation is that some of the installers have to work at great heights on the poles, which support the huge blades that generate energy from the wind. The hazard is multiplied because these workers work under extreme windy conditions in open fields[53].

In addition, both wind and solar systems are so uncomplicated, that they are well suited for homes and small business operations occupying a single building, or a cluster of buildings in relatively close proximity to each other. The advantage of close proximity is that, in transmitting electricity by wire out in the open, up to 80% of it can be lost to the atmosphere it transit from the power source to user buildings, depending on transit distance. These applications are particularly suited to rural operations located off the electric power grid.

The experience in rural India is spectacular. Communities without access to the electric grid have "leapfrogged" into the 21st century on the wings of solar electricity. These rural villages now have access to the worldwide web. However, the star of their solar revolution show has been solar cookers. These are simple box-like utensils lined with reflector material. According to the Worldwatch Institute, over 100,000 in rural areas have taken up the use of solar energy. The demand for solar cookers has reached urban settlements, which has opening up a market for the cookers[54].

India has recognized the value of renewable energy to its population. India's Ministry of Non-conventional Energy Sources has been supportive. As far back as 1982 a commission was set up to study the feasibility of alternative energy sources. Out of this effort, the Ministry was born. It now subsidizes a wide variety of local small businesses, which manufacture and sell appliances and materials related to a budding solar supply industry[55].

Other Popular Energy Alternatives:

A variety of biofuels have been developed to provide substitutes for gasoline. One of the most popular alternatives is ethanol. However, it is one of many alternatives offered to merely compete with the petroleum industry. Earth has provided alternatives that do not need to be burned to produce energy. Energy does not have to produce extra carbon.

The options that sacrifice food or water for energy are counter-productive in our effort to reduce carbon production. Solar is by far the most lucrative choice for global consumption. Furthermore, as we work together across national and economic boundaries to neutralize the pace of global climate change, other countries become less intimidated by market dominance of the US economy. As such, and especially since the Paris Accord of 2015, the newly reclaimed market freedom to proceed in consensus will prevail.

The problem with merely creating an alternative "less disastrous choice is that it also tempts entrepreneurs to seek to turn scarce food into more wasteful fuel. Most of the poorer countries do not have access to fossil fuels. For them solar is the ideal alternative. They cannot afford to exchange food for fuel . . . . . and they do not need to! They all have adequate access to the sun and/or daylight[56].

The sun is such a natural choice, that, once you begin to appreciate how completely Earth and the universe has prepared available energy for us, the value of technology is given its proper perspective. Our learned instincts to trust technology to provide everything is misplaced. We have to view technology in the context of ALL THE POSSIBILITIES that our environment has already given us. Earth has, indeed, provided all we need; not the coal, oil and nuclear industries!

Risk and Sacrifice:

In debates over economic development and jobs it is almost always assumed that creating jobs is important to economic development. In reality, more than a half of jobs created since the early 1990s have been minimum-wage jobs. The minimum-wage throughout most of the US is well below a living wage. In some cases it is less than half a living wage. Such jobs do net bring economic improvement to the respective regions. Workers tend to over-work, (60 to 80 hours per week) to feed their families. That alone compromises basic work safety. Jobs in certain industries are particularly hazardous. Not only does coal mining desecrate the land, (mountaintop removal), the industry has one of the worst records of worker safety hazards in the economy.

Mining in general is notorious for creating water contamination disasters, especially in poor countries. The public cost of accommodating these high-risk industries and the jobs they create can be staggering. In El Salvador, over 90% of the country's drinking water has been contaminated by mining activities. And the country reaps no benefit from those mines.

CHAPTER 19

Water as a Resource

THE ONLY RESOURCE more critical to the survival of animal and human life than fresh water, is clean air. We cannot survive without air for more than a few minutes. Without water, we have a few days to a week.

Yet, throughout the world, where running water comes through taps, people tend to run water, as if there were an infinite supply. There is a sense that one has only to turn on the tap and the water is there.

However, considerable planning and construction have preceded its availability. There are many hidden costs behind the seemingly "free" flowing water. This is especially true for large cities with considerable investment in the following resource ingredients:

- Expensive infrastructure that transports the water to the city
- There are major costs related to storage in reservoirs
- Water testing and treatment incurs ongoing costs
- Distribution of water within the city is expensive
- Maintenance and incidental repair of the distribution grid is also ongoing
- Deferred maintenance inevitably increases operating costs.

The availability of water poses additional limits in many locations. First of all, is the fact that planet Earth has had essentially the same amount of available water for millennia. However, the global population has grown from about 200 million in the year 1 to almost 7 billion in 2011. It was 1 billion around 1804 and doubled to 2 billion about 1927. It doubled again by 1975 to 4 billion.

All this time, Earth has had to share roughly the same amount of fresh water. Earth's weather and climate have long dictated the population distribution throughout human history. Although most of the written history readily available to the modern western world has focused around the Middle East, and Europe, there must have been larger populations in other parts of the world in ancient times. Below is a table that shows the growth of the world population since Year 1 in the Christian era.

World Population Growth	
Year	Population
1	200 million
1000	275 million
1500	450 million
1650	500 million
1750	700 million
1804	1 billion
1850	1.2 billion
1900	1.6 billion
1927	2 billion
1950	2.55 billion
1955	2.8 billion
1960	3 billion
1965	3.3 billion
1970	3.7 billion
1975	4 billion
1980	4.5 billion
1985	4.85 billion
1990	5.3 billion
1995	5.7 billion
1999	6 billion
2006	6.5 billion
2009	6.8 billion
2011	7 billion,

In effect, we are over-populating our global water supply. A "reserve" supply has been stored up underground in aquifers throughout the world. The Ogallala Aquifer of the central plains of North America is one such example, (see discussion below)[57].

We waste too much water in our industrial cities. Our urban lifestyle takes water use, (and therefore, abuse,) to another level. Consider urban vs. rural per capita use; residential vs. industrial use per capita. Household water use distribution[58]:

- US Household Water Use:
- Toilet use   27.6%
- Laundry   21.7%
- Shower   16.8%
- Faucets   15.7%
- Leaks   13.7%
- Bath   1.7%
- Dish washing   1.4%
- Other uses   2.3%

Running out of Water:

The modern industrial urban complex seems oblivious to the fact that we are running out of water! The denial surrounding this reality is greater than the denial of global warming. Eight cities of particular concern are: Tokyo, Miami, London, Cairo, Sao Paulo, Beijing, Bangalore and Mexico City. Each has a different crisis precipitating the pending shortage. However, the common denominator has been an inability to keep up with population growth. Usually antiquates sewer systems are inadequate and leakage always seems to emerge to exacerbate the crisis[59].

In our modern industrial world, the urban bias makes rural living seem insignificant, backward and inefficient. To make matters worse, a recent trend in international assistance to poorer countries has taken this bias to a level, which categorically invalidates rural dignity. In Bolivia a plan was executed to take water from rural populations to supply cities and selling water back to the rural areas to resupply their needs. However, what had been available to them at no cost was replaced with their water sold back to them at city prices. This was the arrangement executed when the International Monetary Fund, (IMF), awarded a loan to Bolivia to privatize the development of the country's central water and sewer distribution[60].

Privatization and the Hidden Agenda:

The loan of $138 million was expected to significantly improve the nation's economy. With the loan, the Bolivian Government negotiated behind closed-doors with Aguas del Tunari, a multinational consortium of private investors, which included a subsidiary of the Bechtel Corporation of the USA. As the sole bidder, they received a $2.5 billion contract to take over Cochabamba's municipal water system.

Water in, on, and above the Earth

Liquid fresh water
Freshwater lakes and rivers

Howard Perlman, USGS
Jack Cook, Adam Nieman
Data: Igor Shiklomanov, 1993

All Earth's water, Liquid fresh water and Freshwater
in lakes and rivers

Spheres show:

(1) All water (sphere over western U.S., 860 miles in diameter)

(2) Fresh liquid water in the ground, lakes, swamps, and rivers (sphere over Kentucky, 169.5 miles in diameter), and

(3) Fresh-water lakes and rivers (sphere over Georgia, 34.9 miles in diameter). Credit: Howard Perlman, USGS; globe illustration by Jack Cook, Woods Hole Oceanographic Institution (©); Adam Nieman.

Here is an example of the **Colonial Legacy,** where back-room deals determine the conditions under which associates connected to the leadership of major multinational entities receive lucrative contracts. Moreover, the oversight and quality control of the project often lacks local expertize and the international body funding the project often fails to monitor actual environmental impacts[61].

Such a maneuver is common in countries, where the local politicians convince their constituents that modern technology is the best possible option for their needs. "On October 11, [1999] Aguas del Tunari officially announces that it has been awarded 40-year concession rights to provide water and sanitation services to the urban residents of Cochabamba"[62]. Such water projects generally involve dams built in rural areas, where the locals live off the land at a subsistence level. Cut off from their accustomed supply, the rural population, to whom the water had always been available at no cost would henceforth have to buy back limited amounts at city prices.

Local farmers on the countryside from which the project's water would come were surprised and angry. With the support of national concerns, they organized a nationwide protest.

The cost of expanding water delivery and to upgrade the city's water infrastructure was to increase rated by 35 percent. However, when water rates, doubled and tripled, Cochabamba protesters, organized a general strike, which shut down the city of La Paz for four days, in January 2000. In February 2000, the protests become violent, when riot police attacked the crowds.

Earth makes water for all of us to drink and enjoy. It is made
clean and fresh. Something, or someone may come along and
contaminate it. But that is not what nature intended.
Earth's water is free! It is not to be **privatized**!

The "War over Water" hit the international press. The protests spread
across the country. In March, an unofficial referendum was conducted
among 50,000 voters, of whom 96% voted to disapprove the privatization of the water. They also complained about high unemployment and a
generally weak economy.

The government refused to terminate the contract, but an April protests would leave six dead and dozens injured and forcibly detained by
authorities[63]. However, by April 12, the government reversed itself, released the detainees and handed over leadership to the rebel coalition.

After many months of disputes and negotiations, in April 2002, Bechtel
officials agree to settle. The settlement with the new President of Bolivia,
Oscar Olivera was signed at the Bechtel headquarters in San Francisco.

Yet, the prevailing attitude throughout the world's urban communities continues to be that big business is the only source of competent
project execution worldwide. Since the 1980s, the World Bank has put
an emphasis on privatization, in extending loans and aid to developing countries. And disasters like the recent mine disaster of November
2015 in Brazil continue to devastate the poor countries.

What tends to get lost in the glamor of the project country being awarded a major loan is the fact that the country is spending a promise. Whether or not the project is successful the taxpayers are stuck with paying back often twice what you borrowed. Even when the project is successful, in and of itself, the long term consequence is usually a fiscal disaster[64].

Take the chronic problem with privatization in the USA. These projects are in effect, designed to add expensive change orders once construction is underway. In this setting, no politician dare pull the plug on the runaway spending. This phenomenon is repeated on most major private contracts bid on public projects today. The Bay Bridge, connecting San Francisco and Oakland in California was eventually built for $6.4 billion, $3.6 billion over budget[65].

The Privatization Dilemma in the US: Throughout the country biding processes are no longer designed to secure the lowest project cost. The successful bidding party is not held to the successful bid. Thus, cost over runs are acceptable and the initial bidding process becomes a mockery.

The country is at a precarious crossroad at present. Throughout the infrastructural network, many critical gaps exist, (see Chapter 22). Those systems responsible for water distribution, supply, quality and overall maintenance require urgent attention at a time, when the elected officials seem to be distracted by personal and political issues unrelated to serving the public. It is not a favorable time to address the environmental dilemmas, which threaten future generations.

Other Barriers to Available Water: Water supplies are insufficient to serve all life on Earth adequately, given current distribution patterns. This dilemma is particularly distressing, because the popular urban myth in most urban settings is that water is free and can be wasted.

Throughout human history many battles have been fought over the availability of water. We in western civilization have taken this struggle to new heights by commodifying certain critical resources, such as water, which should be available for everyone.

Moreover, our political systems have forced certain people, (usually the local poor in many jurisdictions), to live in places and under adverse conditions that deprive them of adequate supplies of clean fresh water. Industrial activities often waste inordinate quantities of water. Some of the wastewater recycled by certain industries is contaminated and further contaminates other sources of fresh, clean water.

Water is a necessity for all living organisms. It is naturally produced and should be available to all. Like the air we breathe, water should not be for sale. Governments in civilized society should manage or facilitate the availability and distribution of water for its constituency. As such, drought-stricken regions should not have water privately owner and/or available for sale[66].

The choice of certain industries impose an amplified drain on available water supplies. From the growing of almonds in semi-arid regions to the mining of coal copious amounts of water are applied to the process of preparation for consumption. Similarly, the choices of using certain human consumer products require the equivalent of thousands of people per item. These include swimming pools and green-grass lawns in arid regions. Their use is amplified in regions where water is scarce[67].

Nevertheless, Earth manages to purify the snowmelt into clean, fresh spring water – "the best you ever tasted" – from a mountain stream. A company bottles it, add a label and you would buy it off a grocery shelf as desirable premium water. Its plastic container often ends up in the local landfill. So much plastic is used for storage and containers of liquid merchandise, that it eventually contaminates the ocean with huge flotillas of discarded plastic waste[68].

The Hidden Human Costs of Mining Activities:

According to an Oxfam study, very few mining operations throughout the world have generated significant wealth to the communities in which the mines are located. This phenomenon is particularly true in Central America, where none of the mining enterprises have played a significant role in the resident country's economy[69].

The general disposition of mining site communities is that they bear the brunt of the human cost of each operation. Mining, especially open-pit excavation leaves scars on the land that take generations to heal. Even when the mining excavation is underground, great volumes of excavated waste is generally piled up in the vicinity and left in the form of relatively unstable hills. In many cases unusual amounts of hazardous materials are left untreated, resulting in extensive contamination of the region's groundwater supply.

Such contamination often enters into major streams and rivers in the region. The experiences reported in the sections on coal mining bear testimony to the cruel fate inherited by the environment in the immediate vicinity of mining operations. In some regions, the entire population is left with a severely compromised drinking water supply. In El Salvador, for instance, mining enterprises were never significantly productive and brought discernable long-term value to the country[70].

Hydro-Electric Power:

Hydro-electric power is a source of non-carbon initiated energy commonly installed in major rivers to serve large metropolitan areas. It generally involves building a dam at a strategic location on the river. Dams have the potential of adversely affecting many local interests. The weight of the water held back by the dam contains the latent energy, which, when released, turns the turbines that generate the electricity.

However, upstream from the dam, considerable farm land and settlements may have to be displaced in order to flood the area proposed to accommodate the reservoir. Downstream a different type of displacement comes, (often more gradually) from the reduced flow of the river. A wide continuously flowing river, can be reduced to a seasonal trickle. People, who relied on fishing for food may be deprived. Irrigation for farming is significantly reduced.

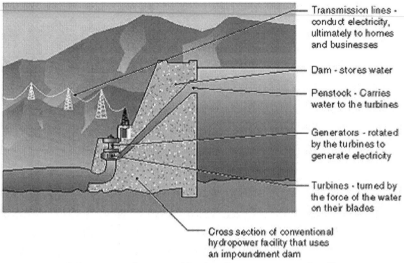

Transmission lines - conduct electricity, ultimately to homes and businesses

Dam - stores water

Penstock - Carries water to the turbines

Generators - rotated by the turbines to generate electricity

Turbines - turned by the force of the water on their blades

Cross section of conventional hydropower facility that uses an impoundment dam

Schematic of a typical hydroelectric power plant[71].

The most common type of hydroelectric power plant is an impoundment facility. Such a facility is typically used for a large hydropower system. As such, river water is stored behind the dam, creating a reservoir. The weight of the water in the **reservoir** provides the power. A **gate** controls its released into the **penstock**, which flows under pressure to turn a **turbine**, which in turn activates a **generator** to produce electricity. The electric power is then distributed by **transmission lines**. Under ideal conditions, the reservoir would maintain a constant reservoir level[72].

Hydroelectric power can be derived from both river and ocean sources. Rivers are by far the most common source of hydro power. However, in strategic parts of the world, the daily changing ocean tides can then be used to generate electricity. Hydropower is renewable, clean and a sustainable source of power.

Hydro power is relatively inexpensive to generate. Power plants have relatively long lives and the amount of power being generated can be regulated at any time. It, along with many other renewable sources, is an energy sources, which, by its use, requires no carbon release.

Most disadvantages to the use of hydropower revolve around the building of dams. A dam floods land upstream and cuts off water downstream. Dams interrupt the flow of sediment to the ocean, reducing beach buildup. The bedrock upon which the reservoir sits must be chosen carefully. A section below is devoted to discussing the pros and cons of dam construction and related risks.

Rivers:

A story of rivers in human history reveals a natural resource beyond compare. As a generator of energy, the river is a primary non-carbon-producing "clean" energy source. Even prior to the industrial revolution, water wheel were used to power mills. They were also used to irrigate large fields, using gravity and networks of canals to perform extensive distribution of water and transportation of goods. Today rivers are used for relatively inexpensive transportation.

In the age of electricity, rivers have been extremely valuable as a location for building dams for both hydro-electric power and as reservoirs to provide cities with a ready supply of water. However, the water behind a dam has often required much sacrifice. Entire communities, both upstream and downstream, often give up many of their critical resources to facilitate dam construction and maintenance.

Rivers are a major source of water for drinking and for irrigation. In some regions, such as Egypt, Southwestern USA, rivers, which have their source in well-watered areas become the primary source of water for desert communities downstream. The Nile and the Colorado rivers are two such rivers.

Some rivers flow into natural lakes, which provide water storage. Some man-made lakes are constructed as reservoirs to serve entire communities. Large lakes, like the ocean feed the atmosphere with moisture, in the form of water-vapor, which condenses to form rain-bearing clouds, the main source of precipitation for some critical regions.

Rivers can often be identify by the row of trees that follow the watercourse. Their other values include their capacity to hold and protect the soil around them from excessive erosion.

Providing Critical Transportation:

Many of the great ancient civilizations were built on rivers, often at a point where bodies of water met, such as at the confluence of a major river and a strategic tributary, or where a major river comes to the ocean. Cairo is on the Nile River, near its mouth on the Mediterranean; Beijing is on the Yangtze; London is on the Thames; New Delhi sits on the Ganges; Manaus is a city on the Amazon, where it is joined by the Rio Negro; St. Louis is at the confluence of the Mississippi and the Missouri rivers. The Ottawa River meets the St. Lawrence at Montreal, New Orleans is built on the Mississippi Delta and the Hudson flows into the Atlantic at New York.

The mouths of rivers may sometimes provide a deep-water harbor. , However, if is carry an enormous discharge, its harbor can be overwhelmed by sediment brought down and deposited at the mouth. A river in an undeveloped region may flow for centuries, without building up sufficient sediment to fill in the harbor. However, once that deep-watered harbor attracts significant urban growth, the resulting build-up of farming, the clearing of forests to build housing and other careless land use practices reduce the viability of the harbor. The ancient city of Ephesus saw its famous harbor reduced and eventually eliminated. The city is now five miles from the Aegean Sea and the bay in which it was located has been completely filled with sediment. At one time it was one of the largest cities of the Ancient Roman Empire[73].

The Dam Impacts:

Every watershed has its own unique ecosystem. Accordingly, the construction of dams and artificial levees along the course of a major river can have far-reaching impacts.

This phenomenon was particularly hard on the natives of the Sierra Nevada Mountains, as the migrants from Eastern USA crossed into California and proceeded to dam the rivers. The damming displaced the native populations both above and below the respective dams. Above the dam their homes were flooded out and below the dam the severely reduced flow of water rendered their fishing grounds useless. Most of the area within the foothills of the California Sierras is semi desert. Farming in this region is virtually impossible without irrigation.

Dams also impact the beaches near the mouth of the dammed river, as the dams starve the ocean of its normal supply of sediment, from which the sand supplied the beaches. Many coastal communities around the USA are experiencing such a phenomenon, in which the beaches begin to disappear.

Lake Mead and other dams on the Colorado have been so exhausted by excessive use for urban development that the river ceased to flow to the ocean in Mexico for many decades. Too many cities are tapping into this limited water supply. The level of Lake Mead is steadily being lowered beyond the capacity of the Colorado to keep up with its consumption.

Put into a more focused perspective, one should note that six of the top ten metropolitan areas in the USA likely to run out of water in the foreseeable future are in the southwest. Four of these six are in the seven-state region, which feeds off the Colorado River[74].

Beginning in the 1920s, Western states began to lay claim to the Colorado's water, building dams and diverting the flow hundreds of miles, to Los Angeles, San Diego, Phoenix and other fast-growing cities. The river now serves 30 million people in seven U.S. states and Mexico, with 70 percent or more of its water siphoned off to irrigate 3.5 million acres of cropland.

Lake Mead, which shares the boundary between Nevada and Arizona, was formed by the Hoover Dam. The lake is 110 miles long behind a 700+ foot Hoover dam face. But the lake's water level is far lower than where it once was—some 130 feet lower, since 2000. Water resource officials say some of the reservoirs fed by the river will never be full again.

The Colorado River's water flowed into Mexico, when the
Morales Dam was opened to receive water being released
for the first time in many decades from Lake Mead.

In effect, so much water is being taken out of the Colorado River
today; that the river had not reaches its natural mouth in the Gulf of
California, in Mexico for decades. The river's most southern dam —
Mexico's Morelos Dam, near Yuma, Ariz. trapped the last trickle of water that got to that part of the Colorado in an attempt to save the "last
gulp" for Mexican farmers, who used to farm in the delta for many miles
north of the Gulf[75].

Most of the traditional Mexican farmland around the Gulf had to
be abandoned for decades, during the imposed drought. The situation
remained a grim testimony to our "improvement on nature" to create
"progress" in our land at the cost of rural survival in that part of Mexico,
until recently. It also stood as a living testimony of a driving force behind the desire for the destitute people at the Mexican border seeking
to migrate to the USA.

Today, only 21 percent of rivers and streams sampled in the USA
were in "good condition". Another 23 percent were in "fair condition,"
according to the EPA survey. The heavy use of chemical fertilizer on
the larger farms reaches the rivers, which drain those farmlands. As a
result, across the nation, 40 percent of rivers and streams have high

nitrogen levels, and 28 percent have high levels of phosphorus, according to the.

A similar EPA survey study found nearly 50 percent of lakes across the country contain fish with levels of mercury that exceed safety limits for human consumption. For most people, the health risk of mercury from eating fish is not a concern unless they regularly eat fish from contaminated waters. The mercury risk is significantly higher for unborn and young children.

High levels of bacteria were found in 9 percent of streams or rivers, making those bodies of water potentially unsafe for swimming and recreation.

A Regional Problem

The southern coastal plains of the southeastern states and the Mississippi River basin had the highest percentage of polluted waters, with 71 percent in poor condition and only 12 percent in good condition. In southern Appalachia, 65 percent of rivers and streams are in poor condition. Fifty-seven percent of streams in Northern Appalachia, which spans from northern Ohio to Maine, are in poor condition, and 59 percent of streams in the northern Midwest are in poor condition.

River and stream health is highest in the northwestern states of Washington, Oregon, Idaho and Montana, where 42 percent of streams are in good condition and only 26 percent are in poor condition.

Time Perspective: It is especially difficult for us, in our modern frame of mind, to realize and keep in the forefront of our minds that we live in a cultural lifestyle fantasy! Much of our time perspective is exclusively based on the artificial framework of clock and calendar. As such, every natural event we consider in clock and calendar terms is to some extent unrealistic. That is why much of the weather, climate and season phenomena which we normally expect, so often surprise us. Our dilemma is further exacerbated by our cultural divorce from the traditions and wisdom of the ancestors, over the past several generations.

Ironically, the traditional societies, which remain on Earth are considered in such low esteem, that most of us are virtually unaware of the potential of learning from them, nor would we anticipate their wisdom. In this age, it remains critical that, wherever possible, we seek to engage those, who are still conscious of ancient ancestors and their wisdom and to learn as much as possible from them before they and their wisdom become extinct, 76.

Notes

1. The global urban society is predominately patterned after domination systems, in which a few people seek to control the majority, who are a created underclass. As such, the minority ruling class has a built-in advantage over a distinct common folk. See: <u>Engaging the Powers</u>: Discernment and Resistance in a World of Domination, Fortress Press, by Walter Wink, Minneapolis, 1992.

2. The greenhouse effect occurs naturally, having a stabling habitable climate. Thus, when the carbon dioxide content is increased the resulting higher temperature is stabilized. The oceans are a critical part of the climate system, with vastly greater thermal capacity than the atmosphere. Most of the net energy increase in the climate system in recent decades is stored in the oceans. A small, slow increase in their temperature is significant. Over one-third of human-induced greenhouse gas emissions come from the burning of fossil fuel to generate electricity.

3. In the chaos that prevailed after the fall of the Roman Empire, people were left without the protection of the Roman soldiers. Thus, they relied on the landowners and their hired security. In effect, the peasants provided most of the food, clothing and other essential materials that supported the feudal world and lived in debt, while providing it. See: Characteristics of the Feudal World, a Public

Broadcasting Station, Copyright©2008, Northeastern Educational Television of Ohio, Inc. All rights reserved.

4. "The patterns of domination and oppression that continue to afflict Indigenous Peoples today throughout the world are found in numerous historical documents such as Papal Bulls, Royal Charters and court rulings. For example, the church documents Dum Diversas (1452) and Romanus Pontifex (1455) called for non-Christian peoples to be invaded, captured, vanquished, subdued, reduced to perpetual slavery and to have their possessions and property seized by Christian monarchs. Collectively, these and other concepts form a paradigm or pattern of domination that is still being used against Indigenous Peoples." These are excerpt from: Statement on the doctrine of discovery and its enduring impact on Indigenous Peoples at the World Council of Churches, Executive Council meeting at Bossey, Switzerland, 14-17 February 2012. The concept of "Manifest Destiny" coined by the USA at least since 1818, when Andrew Jackson led military forces against the Seminoles in the Floridas, reasoning that destiny intended that America should have these lands, anyway. See: "American History" by Michael Lubragge, University of Groningen, c1994-2012, GMW.

5. In this campaign natural stewardship and a love for nature was replaced by plantation agriculture, slave labor and a hatred of both land and laborers.

6. The excerpt from: Statement on the doctrine of discovery and its enduring impact on Indigenous Peoples at the World Council of Churches, Executive Council meeting at Bossey, Switzerland, 14-17 February 2012 have been echoed by many other religious bodies since then, including a Quaker Indian Committee, The Episcopal Church of the USA, and many other bodies. United Nations Declaration on the Rights of Indigenous Peoples in 2012 stated: "Doctrine of Discovery, Used for Centuries to Justify Seizure of

Indigenous Lands, Subjugate Peoples, Must be Repudiated by United Nations." The statement was delivered by the United Nations Economic and Social Council, 8 May 2012. The UN Declaration is not binding. However, its statement illustrates a rising tide of justice being desired by a significant global voice.

7. Owen Jarus explains their system: In exchange for labor, the Inca government was expected to provide feasts to the people as a form of payment, since they had no currency. See: "The Incas: History of Andean Empire", by Owen Jarus, Live Science Contributor, November 19, 2013. Also read: Charles C. Mann: 1491, First Vintage Books, New York, NY., c2006, pp. 81-82, 95-98.

8. Some epidemiologists believe that 90% of the Incas perished in that epidemic, leaving a small, devastated community to defend themselves against the soldiers of the Doctrine of Discovery. Read an abbreviated history of the Incas of that time: "The Greatest Mystery of the Inca Empire was its Strange Economy", by Annalee Newitz, August 27, 2013.

9. Amaranth is one of the crops that enjoyed greater yields among the Incas than what we experience today. Amaranth can be cultivated as a grain crop, or as a vegetable crop. It is high in protein: Amaranth: Modern Prospects for an Ancient Crop, (BOSTID, 1984, 74 p.) Based on work by Chris Watkins, Curt Beckmann and Eric Blazek, 1 July 2011.

10. In his book: Getting to the 21st Century: Voluntary Action and the Global Agenda, by David C. Korten, for the People-Centered Development Forum, 1 January 1990, David Korten explores alternative potential policies that would offset the unsustainable direction of current global development, based on an unrealistic assumption that economic growth could continue indefinitely.

11. These comments were presented during the Workshop: "To Change the Future, Change the Story" at the Guiding Lights Conference, Seattle 2012

12. The concept has been around since prior to World War II. Its potential is yet underdeveloped. Nevertheless, it holds great promise, especially in the production of food within large cities and metropolitan areas. See: Dirt-Free Farming: Will Hydroponics (Finally) Take Off? By Eric Siegel, Illustrations by Jason Novak, June 18, 2013.

13. "They say that the language we are taught prepares us for the society we are going to be living in. Our language prepares us to live in a domination system where a few people (largely hidden from view) will control the large majority." By living in this system and calling it "democracy" some people come to expect far less from their democracy than they should, while the few privileged take for granted what extras they receive by virtue of their social status. Furthermore, they claim to deserve more, because they have much more in society than the poor. Read: "Domination Systems," by Duen His Yen, Copyright c2006 by Duen His Yen, All rights reserved. Last updated 15 February 2007.

14. This figure was taken from: A Brief History of Atmospheric Carbon Dioxide Record-Breaking, by David Middleton for World Meteorological Organization, Press Release #965, December7, 2012.

15. The make-up of today's climate is unlike anything we have documented in recorded time. As such, we can no longer rely on past history to assess current situations, nor can we estimate future developments, based on current evidence. Bill McKibben refers to this new environment as a new planet, which he calls Eaarth: see: Bill McKibben: Eaarth: "Making a life on a tough new planet", St. Martin's Press, New York, NY., c2010, 2011.

16. The make-up of today's climate is unlike anything we have documented in recorded time. As such, we can no longer rely on past history to assess current situations, nor can we estimate future developments, based on current evidence. Bill McKibben refers to this new environment as a new planet, which he calls Eaarth: see: Bill McKibben: <u>Eaarth</u>: "Making a life on a tough new planet", St. Martin's Press, New York, NY., c2010, 2011.

17. The following article describes these flaws in the carbon-offset strategies: "Climate Change 2013: Where We Are Now - Not What You Think", by Bruce Melton, Truthout, News Analysis, 26 December 2013.

18. The flood of skeptical responses to the Paris Accords illustrates the range of sentiments aroused by their success. As the Climate Debate continued it is clear that many cannot imagine a democracy that engages the wide spectrum of opinions captured in the Paris forum. See: Bottom-up Paris Accord May Surprise Activists and Skeptics. By Ronald Bailey, December 9, 2015.

19. One of the first voices of unwavering support for the Paris Accord came from Bill Gates, who was quoted in the Washington Post on Climate Change: "We need to move faster than the energy sector ever has", by Joby Warrick, Energy and Environment, The Washington Post, November 30, 2015.

20. The internet has created unprecedented opportunities for global communication and cooperation, as documented by Paul Hawken in his book <u>Blessed Unrest</u>, How the Largest Movement In the World Came Into Being *and* Why No One Saw it Coming, Viking Press New York, NY., c2007

21. In all, 140 foundations funneled $558 million to almost 100 climate denial organizations from 2003 to 2010. Meanwhile the traceable

cash flow from more traditional sources, such as Koch Industries and ExxonMobil, has disappeared. Powerful funders, he added, are supporting the campaign to deny scientific findings about global warming and raise doubts about the "roots and remedies" of a threat on which the science is clear. Taken from paper titled: "Dark Money" Funds Climate Change Denial Effort by Douglas Fischer and The Daily Climate, 24 December 2013.

22. Dahr Jamail has interviewed NASA emeritus scientist Robert Bindschadler, who worked for 35 years as a glaciologist at NASA Goddard Space Flight Center. He was, at first discouraged by the effect of the denial movement. However, many of his colleagues banded together, encouraged by the overseas scientists to improve research and publish their results as broadly as possible. The military experts continued to plan steps to counteract the true impact of climate change on their respective responsibility. This support of the truth has been effective. See: Dahr Jamail | NASA Scientist Warns of Three to Four-Meter Sea Level Rise by 2200, by Dahr Jamail, Truthout, 13 August 2014.

23. A Drexel University study finds that organizations that deny global warming are funneling a large slice of the donations going to a third-party to promote their unfounded denials. The report is titled: "Dark Money" Funds Climate Change Denial Effort, by Douglas Fischer for the Daily Climate Dec 23, 2013

24. For some people, winning an argument is more valued than self-respect. Fossil fuel burners purchased willing pretenders to put forth false information to confuse the public. This only works in the US, where most high-school graduates have not been taught basic geography. See the impact in: "The American 'allergy' to global warming: Why?" by Charles J. Hanley, Associated Press Special Correspondent, September 23, 2011.

25. For the miners suffering from post-traumatic stress disorder, the road back to working and normal health is still a far-off dream. When miner José Ojeda went back to working underground, he had an almost immediate blackout, an overwhelming sense of anguish and panic. Chilean Miners Fight Their Demons by Jonathan Franklin, The Daily Beast.

26. The outside world did not find out until the 18th day of their entrapment. The mining company had simply failed to complete that routine safety escape route required by Chilian law. Ariel Marinkovic was involved with the investigation. See "Poor safety standards led to Chilean mine disaster", by Ariel Marinkovic. Also, see link: See also link: http://www.globalpost.com/dispatch/chile/100828/mine-safety

27. BBC Report: "On This Day" 21 October 2005: 1966: Coal tip buries children in Aberfan. More than 130 people, mainly children, have been buried by a coal slag heap at Aberfan, near Merthyr Tydfil in Wales. Probably the greater living disaster has been the traditional success of legislators in the US to protect mine operators from their humane responsibility to the miners, who remain the greatest human sacrifice to today's modern lifestyle.

28. See: "How Job-related Illnesses Upended These Families' Lives", by Jamie Smith Hopkins, Jim Morris and Maryam Jameel, Center for Public Integrity, June 29, 2015.

29. Consider the following disasters created on a regular basis by Mountaintop removal practices: FLOODING — Coalfield residents experience drastic increases in flooding following mountaintop removal operations. These floods are often accompanied by mud slides, which can sometimes leave many houses, with their first floor buried in mud. BLASTING — Families and communities near mountaintop removal sites are forced to contend with continual blasting from mining operations 300 feet from their homes, operating 24

hours a day. The impact of blasting frequently cracks walls and foundations, and can send boulders flying hundreds of yards into roads and homes. SLUDGE DAMS represent the greatest threat to nearby communities of any of the impacts of coal mining. Impoundments are notoriously leaky, contaminating drinking water supplies in many communities, and some have failed completely. Reported in the publication: Learn more about mountaintop removal coal mining by Appalachian Voices, 171 Grand Blvd, Boone, NC 28607, c2006.

30. Mining is one of the most hazardous jobs in labor circles, (see Tables 1 and 2, above). Mining companies are notorious worldwide for their abuse of labor. The record of mining hazards and labor casualties tells its own story: U.S. Department of Labor | Mine Safety and Health Administration (MSHA) | 1100 Wilson Boulevard, 21st Floor | Arlington, VA 22209-3939.

31. The Tennessee Valley Authority, (TVA), leaked a huge coal sludge spill into the Emory and Clinch Rivers, which are tributaries of the Tennessee River. Critics of these spills suggest that the TVA has become "a poster child for the failures of self-regulation". The sludge dam breach in Martin County, KY, in 2000, sent more than 300 million gallons of toxic coal sludge into tributaries of the Big Sandy, causing what the EPA called, "The biggest environmental disaster ever east of the Mississippi." Read: "Toxic Coal in Tennessee", by Kelly Hearn for The Nation, February 23, 2009. Kelly Hearn is an investigative reporter whose work has been funded by the Pulitzer Center on Crisis Reporting.

32. A consortium of 14 local, state and regional advocacy groups have worked together for years supporting the victims of mountaintop removal. The effect they have had easing the level of abuse inflicted on the residents of these communities has been minimal. It is a measure of the effectiveness of "democracy" in the region and the country as a whole, in the face of a practice of corporate

entitlement. Learn More about Mountaintop Removal Coal Mining by Appalachian Voices, Boone, NC, ©2006.

33. Ibid, Appalachian Voices: The coal industry maintains such floods are "Acts of God." Researchers at the University of Kentucky recently concluded: "there is a clear risk of increased flooding (greater runoff production and less surface flow detention) following [mountaintop removal and valley fill] operations."

34. Look closely at the photograph in the text and observe the mark left from the flash flood created by the dam break several miles upstream. Compare it with the level of the normal flow of this river and appreciate the scale of devastation represented here. "Brazilian Mine Disaster Releases Dangerous Metals", by Luisa Massarani, 21 November 2015.

35. Contamination of public waterways, the homes and environment of poor communities, entitlement of coal barons and a general disdain for the lives of the miners and their families account for some of the human sacrifices, which make "clean coal" fantasy seem real to whoever wishes to believe in it. See: "Why Is the EPA Sitting on Its Ash?" By Kate Sheppard, for Mother Jones, December 21, 2010. The reference is to the Big Sandy River disaster related to the Tennessee Valley Authority's power plant.

36. Between 2008 and fall 2012, state environmental regulators determined that oil and gas development damaged the water supplies for at least 161 Pennsylvania homes, farms, churches and businesses. The findings in Pennsylvania are significant because they are some of the first official research studies to show confirmed water contamination caused by hydraulic fracturing Contaminated Water Supplies, Health Concerns Accumulate With Fracking Boom in Pennsylvania: by Roger Drouin, Truthout, Wednesday, 26 March 2014

37. Between 2008 and fall 2012, state environmental regulators determined that oil and gas development damaged the water supplies for at least 161 Pennsylvania homes, farms, churches and businesses. The findings in Pennsylvania are significant because they are some of the first official research studies to show confirmed water contamination caused by hydraulic fracturing Contaminated Water Supplies, Health Concerns Accumulate With Fracking Boom in Pennsylvania: by Roger Drouin, Truthout, Wednesday, 26 March 2014

38. The details of these findings are contained in a study: "Earthquake spike pushes Oklahoma to consider tighter fracking regulations", by Peter Moskowitz, for The Guardian, 25 June 2015.

39. The National Parks Service has prepared a description of the inner workings of a geyser: Hydrothermal Systems: a synopsis of how geysers produce jets of super-heated steam, called geysers. Prepared by the National Parks Service, updated December 23, 2015. In California this natural energy producer has been harnessed to produce an operating power plant. See: Calpine: America's Largest Geothermal Energy Producer, by Paul Lester, Calpine, October 6, 2015.

40. Retrieved from http://en.wikipedia.org/wiki/Renewable_energy_in_Iceland Also, Ibid. Christopherson, 5th Ed.c2003,pp.349-352; Lutgens,11th ed.,c2007, pp.139-140; Munroe, et al., 6th ed., c2006, pp. 519-521.

41. The study: <u>World-Wide Direct Uses of Geothermal Energy 2000</u> by John W. Lund and Derek H. Freeston, University of Auckland, New Zealand was presented to the Proceedings of the World Geothermal Congress, 2000.

42. Nuclear reactors are not energy efficient. They produce far more heat than they can possibly use. It takes as much as 500,000 gallons

of water per minute to keep these plants cool. Even then, around two-thirds of the heat is wasted and needs to be spilled into nearby waterways or into the atmosphere. A reactor is like a sports car built to travel 600 miles per hour in a world where the speed limit is 60 mph. To operate it safely, you need to have your foot on the brakes - at all times. And good luck if the brakes fail," said Gar Smith, author of Nuclear Roulette, Chelsea Green Publishing, when interviewed by Mark Karlin, Truthout, Saturday, 29 December 2012.

43. The site is paired with 48 stone or concrete 105-ton markers, etched with warnings in seven languages ranging from English to Navajo as well as human faces contorted into expressions of horror, the massive installation is meant to stand for at least 10,000 years—twice as long as the Egyptian pyramids have survived. However, on 14 February 2014 a leak was detected and the site was shut down, leaving an investigating team to locate the leak. By 18 April, they had experiences two spikes in radiation and the 21 workers tested positive for unsafe radiation: U.S. Department of Energy, WIPP Recovery report, last reported, 7/11/14

44. With its capacity to continue to contaminate its environment for 250,000 years, we have no idea how safe it would be in the long term. Moreover, given the geological history of Earth, we know that underground movement, especially in the Earth's crust is utterly unpredictable. A visit to the Grand Canyon would demonstrate possible movement in adjacent rocks millions of years difference in age. See: Nuclear Waste Documentary published by City University of New York, 2011 http://academic.brooklyn.cuny.edu/physics/sobel/Nucphys/waste.html

45. See Nuclear Roulette: The Truth About the Most Dangerous Energy Source on Earth Saturday, 29 December 2012 by Mark Karlin, Truthout reporting on an interview with Gar Smith, author of Nuclear Roulette, Chelsea Green Publishing.

46. In reality, informing the public would empower them to take their own precautions in a calculated time frame, so as not to need to panic in the face of impending disaster. The same can be true of the outstanding need for the public to be apprised of all the "natural disasters" which arise out of our public failure to be proper stewards of Earth. Accordingly, more people may engage in their own disaster preparedness, especially where urban and other landscape designs have failed to take into account the likelihood of natural events having disastrous impact on vulnerable and unsafe development. Furthermore, why shouldn't we be informed of the vulnerability designed for people in the future, who were planning to move into vulnerable locations?

47. At this point in the discussion, it would be helpful to consider a few details related to these operational accidents. Three Army technicians were killed in an explosion at a government reactor in Idaho in 1961 (their bodies had to be buried in lead-lined coffins). Another eight workers were killed in a series of three explosions over a 14-year span at the Surry reactor in Virginia, Nuclear Roulette, by Gar Smith, Chelsea Green Publishing, Vermont, c2012.

48. Concerns about the inherent dangers of nuclear energy generation are significantly magnified by Japan's March 11, 2011 earthquake and tsunami that destroyed safety systems and caused the meltdowns of three of the units at the Fukushima Daiichi nuclear power plant. Published in the Friends of the Earth journal: Beyond Nuclear

49. A 1994 paper by Lynne Page Snyder of the University of Pennsylvania titled, "The Death-Dealing Smog Over Donora, Pennsylvania: Industrial Air Pollution, Public Health Policy and the Politics of Expertise, 1948-1949," published in the Spring issue of the journal Environmental History Review, describes the event: "Pollution from the Donora Zinc Works smelting operation and other sources containing sulfur, carbon monoxide and heavy metal dusts, was

trapped by weather conditions in the narrow river valley in and around Donora and neighboring Webster. Air pollution problems were recognized from the facility as early as 1918, when the plant owner paid off the legal claims for causing pollution that affected the health of nearby residents. In the 1920s, residents and farmers in Webster took legal action again against the company for loss of crops and livestock. Regular sampling of the air was begun in 1926 and stopped in 1935."

50. "Historic Marker Commemorates Donora Smog Tragedy" by David Hess, Secretary, Department of Environmental Protection contributed some of the above information.

51. There are 16 "Fukushima-style" reactors currently installed on 16 sites in 12 US states. When Superstorm Sandy hit the East Coast, it knocked out five reactors in its path - including three GE Fukushima-style reactors. The Nine Mile Point reactor was shut down, the Fitzpatrick reactor caught fire, and flooding at the Oyster Creek reactor came within six inches of disabling the spent fuel pool cooling pumps. (If these had failed, the NRC's recommended "fix" was to use a "fire hose" to cool the plant.) – Ibid. Gar Smith interview with Mark Karlin, Truthout on Saturday, 29 December 2012.

52. The Nuclear Industry does not know the actual safety level of any one nuclear plant. When a meltdown occurs, it can quickly create an international hazard as nuclear fallout spreads. The Chernobyl disaster contaminated the air as far away as Sweden. It is still unclear as to the duration of continued fallout into the future. Then there is the element of trust that can be counted on, when the official tendency is to downplay the danger, so not to risk public panic. It is like the child spilling gasoline in the basement and not alerting the parents, who are smokers for fear of potential punishment. See

Spent Nuclear Fuel: "A Trash Heap Deadly for 250,000 Years or a Renewable Energy Source?" by David Biello, Jan 28, 2009.

53. Both of these studies outline the costs and benefits of wind power: "Environmental Impacts of Wind Power": a study prepared by the Union of Concerned Scientists, Science for a Healthier and Safer World, Also: Capturing the Wind: Power for the 21st Century, by Ethan Goffman, June 2008

54. This advance in rural development is reported in a paper: "Solar Power Reaches 100,000 in Rural India", by the Worldwide Institute, Washington D.C. 20036, Updated 10 December 2014. All Rights Reserved.

55. "Introduction to India's Energy and Proposed Rural Solar-PV Electrification" by Najib Altawell. E-mail: n.altawell@dundee.ac.uk University of Dundee, Carnegie Building, Dundee DD1 4HN, Scotland, UK.

56. The Renewable Fuel Standard (RFS) ensures that those choices at the pump are available to Americans. With the RFS opening up the fuel market to new fuel sources, the renewable fuel industry has been able to deliver economic, national security and environmental benefits, such as: Lowering gas prices by $1.09 per gallon; reducing the US' reliance on foreign oil; supporting hundreds of thousands of jobs; reducing harmful greenhouse gas emissions; etc. © 2014 Fuels America. In contrast, solar and wind energy eliminates the ongoing cost of fuel purchase, creates far more jobs, over the complete range of job needs, reduces harmful greenhouse gas emissions to zero and eliminates both cost and expensive risk of both mining and transporting fuel and the related oil spills and their environmental disasters.

57. The table above is derived from a 1999 United Nations estimate, plus more recent updates.

58. The table below has been compiled from data contained in: "How We Use Water In These United States", prepared by the Environmental Protection Agency, January 27, 2004 http://www. epa.gov/watrhome/you/chap1.html

59. For decades, the conventional wisdom of the industrial countries was to develop the ambitious World III countries to enhance the economies of the respective industrial country. It was always assumed that the poorer country would also benefit. However such local benefits did not occur in many cases. The following article from the New Yorker captures these customs in: "Leasing the Rain", by William Finnigan, for The New Yorker, April 2000

60. This article details the struggle between the private foreign investors and the country, which would incur tremendous expense through a loan, which the people would have to cover. In incurring this prohibitive cost, the government was not sufficiently in touch with the lives of the people, whose water was being stolen to meet the urban needs of the big cities. The poor farmers lived outside the money economy and had never had to buy their own water. Their entire landscape was about to be altered in ways that would force them into a completely different lifestyle. Read: "Bechtel vs Bolivia: Details of the Case and the Campaign", written by The Democracy Center, a fiscally sponsored project of Community Initiatives. © All Rights Reserved.

61. Ibid, William Finnigan.

62. Ibid, William Finnigan and ibid, Bechtel vs. Bolivia.

63. Susan George describes in full detail case studies of actual poor countries and their experience handling a development loan. See: Susan George: A Fate Worse Than Debt, ©1968, pp, 171-188.

64. The "privatization" process has long been deregulated in the US. Accordingly, whatever standards may be implied, the actual process is open to fraudulent practices, which often result in the taxpayer being stuck with cost over-runs, sometimes double the initially quoted project cost, as in the Bay Bridge project cost above. For instance, "In March, Fluor's senior vice president Richard Fierce bragged that his company was saving taxpayers $1.7 billion on the new bridge across the Hudson until one congressman offhandedly remarked that he'd heard the Tappan Zee project would cost $5 billion, not $3.1 billion as the contractor had claimed". This was a quote from the paper: "Innovations or Hucksterism? Three Little-Known Infrastructure Privatization Problems", by Ellen Dannin, Truthout, News Analysis, Monday, 22 December 2014

65. Ibid, Ellen Dannin.

66. Ibid, Ellen Dannin.

67. Although it is a small portion of the State's water is under commercial ownership, no private company should be allowed to take the people's water, bottle it and sell it back to them, as described in: Drought Turns Californians Against Water Bottling Companies, by Katie Lobosco, for CNN, May 26, 2015. In fact, we all drink recycled water. We all contribute to the water in the atmosphere. It should be available to everyone and all living things.

68. Plastics pollution in the oceans has become an alarming problem, particularly for aquatic life, of which we know very little. See: "There's a Horrifying Amount of Plastic in the Ocean. This Chart

Shows Who's to Blame". By Tim McDonnell, for Mother Jones Magazine, February 13, 2015.

69. El Salvador has been struggling to save its water. An estimated 90% of the country's surface water is believed to be "heavily contaminated". Twenty-five percent of the rural population have no access to potable water. One river contained 9 times the acceptable level of cyanide, and 1,000 times the acceptable level of iron. Taken from a report: "El Salvador Fights to protect Water from Mining Contamination", by Delice Williams, The Guardian, Mining Watch, June, 2013

70. What they almost always contribute is excessive water pollution and other environmental hazards. See: Oxfam America: Metal Mining and Sustainable Development in Central America, by Oxfam America Inc., © 2008

71. The indented diagram in the text was copied from the following study: "Types of Hydropower Plants", by Office of Energy Efficiency and Renewable Energy, Environmental Protection Agency. Energy. gov

72. The diagram's schematic lists and briefly describes the critical components of a hydroelectric power plant: "Six Important Components of Hydroelectric Power Plants", by Haresh Khemani and Lamar Stonecypher, Bright Hub Engineering, June 5, 2013.

73. Bible History Online caries maps of the Ancient Roman Empire, which show Ephesus, when it was Rome's second largest city. Map of the Roman Empire – Ephesus, (map and text) Ephesus – (N-6 on map) Taken from: Bible History Online. Today's ruins of Ephesus tells a story of land abuse in a slave society during the beginning of the Common Era. In the typical lifestyle of an affluent slave society

oblivious to the need for stewardship of the land, they failed to share their resources equitably. They failed to appreciate the need to value and protect their harbor as the unique resource that is was. Their huge fertile valley was overworked. As a result, the harbor silted over and became useless. As the harbor became useless, the city was ruined and its economy collapsed. The abandoned farms and agricultural fields left the cleared vacant lands to be eroded extensively. Eventually the entire bay protecting the original harbor to completely fill in. Today, the site of the harbor is dry land, five miles from the present coast!

74. The ten cities expecting to run out of water are described in Appendix J. Four of them: Tucson, Las Vegas, Phoenix and Los Angeles receive a significant portion of their water from the Colorado River. Reported by: "The Ten Biggest American Cities That Are Running Out Of Water" By Charles B. Stockdale, Michael B. Sauter, Douglas A. McIntyre, November 1, 2010

75. The water for the pulse feeding the Mexico dry delta isn't coming at the expense of US consumers and farmers. Its water from Lake Mead that Mexico banked there as part of its own allotment. According to this Outside Magazine online article, the water being restored to Mexico is less than 1 percent of the river's average annual flow: "An unusual sight: Water flows in Mexico's Colorado River", by Tim Johnson, McClatchy, April 1, 2014.

IV

❖ ❖ ❖

Quest for Unlimited Growth

Conflict between Individual Wealth and Community Development in Seeking a Balanced Economy

UNLIMITED GROWTH IS not a sustainable long-term possibility. This is because economic growth depends on a constant flow of resources and opportunities. On Earth, resources renew themselves and are constantly available only where their extraction continues within the natural rate of replenishment. In effect, sustainable growth is a function of sustainable consumption of resources.

A balanced economy, in turn, depends on sustainability. As such, economic balance in a true democracy requires that all the people have reasonable access to the resources necessary for human survival. In fairness to all concerned, those who work to add value to the economy should be afforded reasonable compensation for the work they contribute.

In traditional society it was understood and assumed that everyone deserves to live a life of meaning and purpose. This is the foundation of basic human dignity, freedom and justice, which persuaded people to "live and let live".

The modern lifestyle is based on established inequality. Such a disposition may be convenient for the ruling aristocracy. However, such convenience depends on the human sacrifice of others, especially the underclass. Such human sacrifice leads to human relations, which are routinely in conflict with the very sanctity of humanity.

This is not a foundation, upon which authentic democracy can be built. In fact, the sacrifice of the underclass was and continues to be unnecessary. Today's post-colonial global economy and the political truce that supports it remains unsettled. Too many crises continue to brew, leaving too many people living as refugees in places, where they are unwanted.

Is the work they do, the service they contribute unworthy of a living wage?

The unstable and unsustainable condition creates and maintains an unbalanced economy, which has to be held in place by force, no matter how gentile and proper it may seem. The role of the domination system is to orchestrate the perpetual control of the unbalanced mechanism, which governs most of today's modern society.

The system is at odds with much of what the US Declaration of Independence implies and promises. The Declaration criticized inconsistency in the colonial system. However, it recreated a slightly different version of a similar colonial system within the USA.

This surviving inconsistency infects the core of our daily lives, holding the society in chronic conflict. Our domination system can "keep a lid on it". Such a condition may pretend that "all is well"; but it isn't! In fact, this inconsistency creates and sustains chronic poverty, in the presence of scandalous wealth.

Chronic Poverty:

European civilization was built on a foundation of feudalism, which has never been reconciled. The resulting colonial economy is designed to create poverty. In purely economic terms, this phenomenon stands in the way of our very QUEST FOR UNLIMITED GROWTH, as a perpetual contradiction unto itself:

1. The maintenance of widespread, chronic poverty stifles economic growth in many ways.
2. The process condemns millions of families and households to perpetual poverty.
3. In feudal social terms, the despised poor are actually a critical working support, without which the society so constituted could not function.
4. Such an economy has created and sustained a visible class of outcasts within conservative society.
5. The poor stand out, leaving them vulnerable to continuous profiling.
6. The wealth generated by their suppressed wages provide much of the growth in today's economy.

7. This necessary support of today's economy makes chronic poverty a "politically correct" hypocrisy.
8. In many states poverty is a crime, punishable by law.
9. Homelessness is an even greater crime.
10. Children growing up in poverty have a greater chance of themselves remaining chronically poor in adulthood.
11. Most of the poor are actually hired into legitimate jobs.
12. The legal minimum wage in most states is below a practical living wage.
13. The poor are some of the most routinely and legally abused people in society.
14. The poor participate least in an economy, making them a human sacrifice that is taken for granted.

None of the factors listed above is necessary for a balanced economy. A balanced economy allows most of its participants to play a significant role in its activities. An economy thrives on open exchange. The best staple goods and services are affordable to most people, preserving their dignity. Thus, their life derives meaning from visible participation and contribution to the overall prosperity of the community.

In other words, a truly humane community is open! Its foundation is underlined by a genuine desire for cooperative mutual co-existence. This is the essence of the African proverb of the village raising the child.

The village community encourages full participation on all levels. It isn't that the village community lifestyle is perfect. It is not. However, it has built-in customs that acknowledge its fallibility, promote accountability, and encourage the correction of problems, as they are discovered. In contrast, the domination system is not good at acknowledging its errors, even when they are discovered. Chapter 24, "The Crumbling Infrastructure", outlines a vivid example of this shortcoming, in the process of examining the plight of the neglected infrastructure.

Creation of Poverty: Poverty is not an accident that merely "happens". Poverty is usually the product of a variety of historical, social, political, economic and/or traditional circumstances, which, over time have promoted opportunity inequality. The most common cause of chronic poverty is actually depressed wages.

The poverty-stricken are forced to live under conditions of terminal chronic need. Most of them are the "working poor". In the industrial world, it is customary to designate types of labor worthy of extreme exploitation, the modern version of slave labor. When someone is hired and paid less than a living wage then that person is being forced into the slavery of chronic poverty.

In effect, poverty is being created in the workplace. Furthermore, when specific laws are passed to keep compensation frozen indefinitely below a living wage level, such laws are punitive. This is the case with restaurant workers in the US[1]. The $2.13-per-hour wage has been frozen at the federal level since 1996! The federal minimum wage is currently at $7.25 per hour is actually well below a living wage[2].

Where that worker's compensation is reduced to a small fraction of the value added by one's labor, the worker's value is being suppressed and is being underpaid. The worker has earned a reasonable proportion of that value added by his contribution on the job. The same multinational enterprises often pay workers in other countries far more for the same performance. This is particularly true where unions are allowed to represent the workers' rights and value[3]. The wages paid should represent a living wage. Too many jobs in the global economy are designed to create and sustain poverty in this way.

In today's modern economy, the formula for "cheap labor" creates most of the poverty that exists among us. The "successful" employers of the Industrial Era have often thrived on cheap labor created by under-paying their employees. Such under-payment often represents a prominent part of the employer's profit-margin[4].

Poverty-profit on a large scale has been the hallmark of the industrial revolution. It took an aggressive labor union movement in the 19th century to protect workers from such uncontrolled abuse. Today, we have a legislated minimum wage scale, which is still set below a living wage, with no capacity to automatically keep wages abreast of the rising cost of living.

The one thing all poor countries have in common is a wide disparity in income. What is eminently possible is the promise that the entire global population can be fed by Mother Earth. The barriers to this possibility are formidable. Yet, the task is humanly achievable. It will, however, require a different leadership in the global economies to get us there.

CHAPTER 20

Free Market Enterprise is not Free

SUSTAINABILITY IS FUNDAMENTAL to Earth's lifestyle. Sustainability is also vital to the support of a long-lasting strong economy. A sustainable use of Earth's resources facilitates our capacity to live in the harmony of the natural abundance of Earth's lifestyle. Consider the continuous inter-relatedness of the four cycles.

Similarly, a sustainable economy is one that is balanced by a system of exchanging goods and services with equity. It is a system which supports the exchange of value in ways that provide fair compensation for supporting labor. A fair exchange is central to the spirit of democracy. The domination system is not interested in universal fairness. In the context of colonialism, the domination systems that supported the custom is specifically designed to transfer wealth to the colonial power in question.

The colonial system thrives on the underclass it created and maintained to serve the "mother country". In Europe, into the 20th century women and children were considered little more than property. In the US and much of Europe, universal suffrage was only attained in the 1920s, after an extended struggle, which had begun well before 1832, when the first Parliamentary Reform Act in Britain gave voting rights to male heads of household with property worth over £10 per year[5].

In effect, when the Declaration of Independence of the US was written, women were not considered equal. In fact, neither were men, who did not own a certain amount of property. One comes to appreciate how well established was the concept of exclusive freedom and liberty. The elite male was the only one privileged to pursue happiness.

Thus, as we take for granted the standards of universal human civility, it is helpful to appreciate that many people with conservative sentiments continue to struggle to re-establish the exclusive privileges held sacred by the domination systems of the past. In their mind, high unemployment means they "have the right to" bid down the cost of labor and underpay workers. This phenomenon is well demonstrated in figure 1, below, where the loss of middle-wage jobs is replaced with low-wage jobs in the US economic recovery of 2008.

Since the income disparity is widespread, this imbalance is a detriment to the global economy. A study completed by the United Nations illustrates this result well[6].

In the wake of the Wall Street Market collapse of 2007, the collapse was considered to be primarily a financial problem. So, the banks were given the bail-out money to fix the problem. Their corporate minds saw it as their money. The first thing they did was award the CEOs million dollar bonuses. The "bailout" was supposed to ease the crisis.

In the three years after the recovery officially began 95 percent of the increase in incomes went to the upper one percent of the population. For most Americans, there has been no recovery.

The stimulus package was created by the administration to prop up the failing banks and give them the funds necessary to boost the staggering economy. The stimulus would be facilitated by giving loans to local businesses to hire the people being laid off due to the recession. The banks invested in corporate bonuses and investments overseas, where they could get better returns for themselves[7].

A Case for Equity:

2.2% of Middle-Income Jobs Recovered: The recovery was to have begun around 2010. However, the bail-out money that was supposed to have gone to help local businesses be revived and start hiring seems to have gone abroad to expand offshore investments.

In the first two years of the recovery, the middle class has been decimated, (see table above). Between 2008 and 2010, 6.2 million jobs were lost, of which 3.8 million (61%) were middle income jobs. During the next two years, the total job gain was merely 3.28 million, less than all the middle-wage jobs lost is the previous two years.

Table 1 (ftnt. [8])

Minimum-Wage (and Below) Workers by Occupation, 2013

OCCUPATIONAL GROUP	WORKFORCE
Food preparation and serving related occupations	1,540,000
Sales and related occupations	477,000
Personal care and service occupations	228,000
Office and administrative support occupations	196,000
Building and grounds cleaning and maintenance occupations	183,000
Transportation and material moving occupations	171,000
Professional and related occupations	119,000
Production occupations	105,000
Healthcare support occupations	87,000
Protective service occupations	61,000
Management, business, and financial operations occupations	38,000
Construction and extraction occupations	38,000
Farming, fishing and forestry occupations	35,000
Installation, maintenance and repair occupations	21,000

Source: Bureau of Labor Statistics

PEW RESEARCH CENTER

From another vantage point, only about 730,000 of the jobs, (2.2%), gained were middle wage jobs, (see figure below). It is highly likely that many of the 1.9 million low-wage jobs, (58% of total), added between 2010 and 2012 were taken by former middle-wage earners[9].

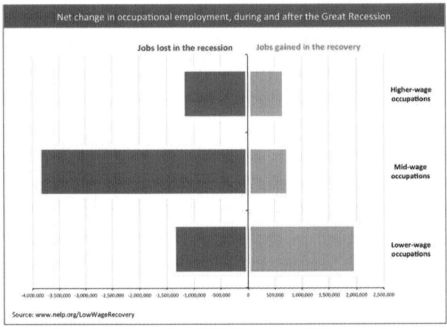

Figure 1. Job Losses from 2008, 1st quarter to 2010, 1st quarter and Gains from 2010 to 2012. Total job loss 2008 -2010 was 6,200,000. Total job gain 2010 -2012 was 3,280,000.

In an effort to create more equitable communities in a few cities, local community development has focused on revitalizing economically stressed neighborhoods. Many struggling downtown business communities depend heavily on active partnership among all stakeholders. Such a partnership thrives on the cooperation between the local businesses and the surrounding residential community, which determines its level of eventual success. The following examples provided by Chattanooga, Tennessee, Portland, Oregon, and Suisun City, California bear witness to inner-city communities that have risen out of the ashes to become more healthy economic centers.

Alternative Scenarios to Urban Development:

In each case, advocates interested in saving the community from economic collapse convinced the local government to invest in extensive community economic development, based on the renewal of a blighted section of the city.

Chattanooga is an intriguing city. It has a few distinct histories, two relate to the ecological history of the area. The old city of the late 19th century was devastated by a series of floods when the Tennessee River reached flood stage and exceeded it by over 20 feet. In effect, the city was initially built in the flood plain of the lower Tennessee River.

The other history tells of the plight of the city during the turbulent times of coal mining. During the national war effort of the early 1940s, the city was important for building naval vessels. However, in the post war years coal mining had brought heavy industry and the chemical industry to the valley.

By the 1950s, the pollution in Chattanooga matched its gritty industrial image. Sometimes the pollution was so thick that street lights had to be turned on at noon. The Tennessee River curled through an industrial no-man's land along the downtown riverfront. Its in-town tributary, Chattanooga Creek, was so polluted from toxic dumping by coke foundries and chemical factories that, in 1994, the EPA proposed 2.5 miles of the creek as a Superfund site.

Finally, Chattagnoogans got fed up. "It was so bad that people couldn't stand it anymore," says Karen Hundt, an urban designer who has been involved in the city's turnaround since those dark days in the 1960s when headlights were sometimes required at noon. It was just that bad[10].

In the midst of the drive to reduce the pollution, the larger plan to revitalize the city began. In that effort, an abandoned bridge over a railroad track was improved and strengthened as a pedestrian bridge, which linked the business districts on either side of the railroad line in the downtown area.

Chattanooga architect Stroud Watson, in taking charge of a new initiative, gave a generation of residents ideas for the design, construction and renovation of what are now city landmarks. His greatest triumphs came through his work at the Urban Design Studio, which promoted the concept that downtown should be a living room made up of districts filled with defined edges and streetscapes. Chattanooga's urban renaissance, founded in 1981 on Vine Street, surged over three decades, through several recessions and recoveries. This effort was coordinated with a major redevelopment project which brought in a museum and other activities, attracting significant tourism into the city, and revitalized the downtown[11].

In **Portland**, parts of the central city were in advanced decay. By the early 1970s, new suburban shopping malls in the neighboring cities of Beaverton, Tigard and Gresham had drawn from Portland's downtown. Unlike many of the country's revitalization efforts at this time, Portland's plan did not call for widespread demolition and reconstruction. The revitalization took on several phases.

The creation of a downtown transit mall in 1977 was the first achievement. Soon thereafter, a new waterfront park in 1978 replaced a section of a freeway. The creation of the Pioneer Courtyard Square in 1984 really spruced up the City Center, which was further enhanced by the opening of the Portland-Gresham light rail line in 1986. The opening of Pioneer Place mall in 1990 successfully retained businesses and drew other enterprises, all of which lured more customers.

When light rail was introduced into the downtown, it allowed a significant section to be turned into a pedestrian mall. Parking problems were eliminated. Transportation into the city center from outlying districts and some suburbs was improved, especially for those with limited transportation options. The significance of the city center was re-established[12].

Suisun City is located about halfway between San Francisco and Sacramento, California. The city had grown from just 2,900 in 1970 to over 27,000 by 1988. Nearly all of that growth, however, has been on the edges of the city, in a typical bedroom community-sprawl pattern.

Meanwhile, Suisun City's Old Town and city center became areas of disinvestment, as industrial oil facilities along the harbor polluted the ground and water, businesses failed on Main Street, about half the City's police resources were used to fight drug dealers and high crime in one neighborhood, and the City Hall's once-temporary trailer housing became permanent.

By 1988, the situation was so bad that a San Francisco Chronicle survey of the quality of life in the Bay Area's 98 municipalities ranked Suisun City dead last. But a growing group of citizens and a frustrated City Council, led by Mayor Jim Spering who was elected in 1989, launched a redevelopment plan to revitalize the City's Old Town and the adjacent downtown areas. The city's Redevelopment Director, Camran Nojoomi, applied the State redevelopment options to merge Housing and Planning with the Redevelopment Agency.

In 1991 the city purchased the dilapidated properties along the waterfront and improved the infrastructure throughout the downtown area. That effort created a new, modern city center, which attracted new entrepreneurs. The expanded new Town Plaza revitalized the entire city, especially the business district[13].

Each of these examples represents a type of urban stewardship undertaken in the spirit of the land stewardship described in this writing. In essence, when we endeavor to take care of our investments, they last longer and can become eminently sustainable.

As shown above, the efforts of Portland, Chattanooga and Suisun City excited the human spirit of their respective communities, which awakened the natural neighborly cooperation within each local marketplace.

A truly free market is evident in each of the above examples. In each case, all participants have almost equal, (and definitely open), access to the local opportunities to share in community revitalization. In each of the three cities, the local community was successfully engaged in the improvement process. Furthermore, each city experienced an increased exchange of available resources in the respective regional marketplaces. A typical community spirit is one that is committed to providing and supporting equal access to that community's resources.

The industrial revolution, in contrast, was built by large merchant industrialists, intent on expanding their personal wealth and competitive advantage in the marketplace. In that context, they championed the ethic of single-minded corporate greed, in which, faithful to the domination system, the ultimate goal was to ruthlessly exploit the community for whatever wealth they could extract from it.

The Doctrine of Discovery Opened the Colonial Era:

In 1452, Pope Nicholas V, as the Papal Head of the Holy Roman Empire, called on the monarchs of Europe to attack the people of the Americas and Africa as "enemies of God", because they were not Christian, (by the Pope's definition)[14]. Today's post-colonial "world order" is a direct product of that movement.

Papal bulls from the 1400s led to an international norm called the Doctrine of Discovery, which dehumanized non-Christians, condoned the conquest of the Americas and other lands inhabited by indigenous people and legitimized their suppression by nations around the world, including the United States. In fact, many of the laws currently being enforced as US law are based on this standard[15].

The Doctrine in the context of contemporary feudal customs and practices: It is helpful to appreciate how normal the Doctrine would have seemed at the time. Christianity was considered the de facto state religion in most of the European countries. The Holy Roman Empire was the seat of Christian leadership. The Pope was the head of the Roman Church. In kingdoms across Europe, the king was considered the head of the church there.

Moreover, it was generally believed that the kings had a divine right to be king. Combined with the station from which the Doctrine of Discovery originated, the attack on the Americas was like the Crusades of a few centuries earlier. It was a religious war against people, who had no idea they were under siege.

The feudal system is a domination system, built on established inequities. It is an unnatural system, so much so that it requires a system of force and intimidation for its survival. Earth's lifestyle is a natural process. It is ultimately in charge of all life on Earth. It is not a system of force. Instead, it exemplifies cooperation. It is so natural, that all, who practise Earth stewardship learn the essential lifestyle of cooperation.

Traditional societies become aware of the harmony contained in the stewardship exercise. That is why their lifestyle tends toward a democratic way of life, complete with its appetite for cooperation and community building.

Because the feudal system is a vehicle for force and intimidation, it breeds violence at its very core. As such, it nurtures violence, which becomes a vicious cycle. We live in such a system. That is why our most common default, when confronted with problems is to "fight"! In effect, our fundamental difficulty addressing the challenges of global climate disruption lies in the way we customarily address challenges.

The Conquest of the Americas: European history of the Middle Ages include dramatic tales of knights in shining armor. This was true of the early crusades. However, getting to the Americas a few centuries later was an arduous trip in barely sea-worthy vessels. The crew on these vessels were the underclass, (often indentured servants), of the post feudal system of Europe. Many were employed by their debtors to confiscate the land the latter would earn as their service in the execution of the Doctrine of Discovery. The writing and journals of Bartolome de las Casas describes the ordeal imposed on the natives of Hispaniola and other parts of the Americas[16].

Bartolome de las Casas was a Dominican priest from a farming background in Seville, Spain. In 1530 he began writing a Latin treatise, which became one of the most significant missionary tracts in the history of the Church. At first, he was attracted to the adventure of travelling to the Americas, impressed with the aura of Columbus. However, as he saw first-hand the real impact of the Doctrine: to "invade, search out, capture, vanquish, and subdue" the people in new lands, to "reduce their persons to perpetual slavery, and to apply and appropriate to himself and his successors the kingdoms, dukedoms, counties, principalities, dominions, possessions and goods, and to convert them to his and their use and profit," he was appalled.

Working with two bishops in Guatemala, he developed the landmark papal bull, *Sublimis Deus*, often called the Magna Carta of Indians rights. In 1537 he forwarded to Pope Paul III a petition that the Indians were truly human and capable of receiving the faith and that they were not to be deprived of their liberty or property, even though they may be outside of the faith[17].

As indicated earlier, the feudal system of 15th to 17th century Europe did not encourage Earth stewardship. In fact, a generally negative attitude underlined the prevailing sense that only slaves worked the land, and only because they were forced to. The period of slavery may have intensified the negative attitude toward the land. Much of that attitude persists in today's modern society.

By the time European colonists began arriving, the native population had been decimated by the European diseases[18]. In places the land was virtually empty. However, the attitude toward the natives had already been set by the spirit of the Doctrine of Discovery and remains intact today among much of the North American European population.

Not only did many of the religious European immigrants believe that the death of the natives was a gift from God, justifying the Doctrine in their minds, it also gave further conviction that their settlement and occupation of the Americas was "manifest destiny". Thus, upon landing in the Americas, those on the lowest level of feudal Europe earned a status elevation. They were now a class above the slaves. They could also claim the basic benefits of "manifest destiny". Accordingly, many feel deeply convinced that North America should be and remain exclusively European[19].

Manifest Destiny and Entitlement: It is very difficult for many, who believe in manifest destiny to take seriously the prospect of anthropogenic climate disruption, (ACD). Raised in the religious environment, which worshipped the concepts of the Doctrine and impressed by the "American Dream" in the context of "manifest destiny", the pride of American Exceptionalism emerged. This sense of pride, entitlement, and invincibility fueled a rampage across the Americas, which completely displaced the natives and made them foreigners in their own land[20].

Then came the Trans-Atlantic Slave Trade and a new form of feudalism. These were African slaves, a superior stock of serfs, who would be bred exclusively for a future of perpetual slavery. It was not long before the plantation owners began to reap an economic bonanza.

The slave trade and slavery were major economic drivers. They could do the work in the field at a lower cost than their indentured servants brought from Europe and they could hire the former serfs to over sea the slaves. In fact, this was a welcome promotion for the former serfs.

The Economic Development of Europe: First of all, it should be appreciated that the economic development of Europe by the exploitation OF America's wealth was wasteful in purely economic terms. Consider the economic value of the bison fur trade. They wiped out a herd of millions of bison in a couple decades to build their short-lived bison trade. The trade died a few decades later.

Had they been disposed to practise stewardship of their resource, the bison herd, they would have managed the herd to their long-term benefit. However, the mentality and habits of "manifest destiny" encourage the habits of affluence and its cousin wastefulness, to champion the customs of the aristocracy, which abhors sustainability and eventually destroys the nest that supported their development. Sometimes, they destroy the nest before they learn to fly!

In this model of economic development, much of the growth of Europe was fueled by the unnecessary impoverishment of the respective natives. They often failed to appreciate that had they been more efficient in their economic development, bringing the native economies along in their growth, they would have also developed a better, stronger market for their eventual products. However, this level of development requires a serious long-term approach.

<u>What Manifest Destiny Fails to Teach us</u>: The attitude of entitlement in the spirit of aristocracy cancels out the opportunity to learn from our inferiors. The arrogance of the Doctrine of Discovery runs so deep, that the Europeans did not consider that their hosts had learned to live in this physical environment, which was new to the Europeans. The failure of the Jamestown settlement of 1595 was largely due to their failure to observe how the natives lived and how they handled winters.

As discussed in Section II, the natives lived *IN* the forests, which reduced the severity of winters. The homes built by the natives were also much better insulated. These benefits come with a lifestyle of observation of the environment, a stewardship habit.

The Long-term Value of Fair Exchange:

Groups like The Rotary, Kiwanis, Soroptimists, Better Business Bureaus and some local Chambers of Commerce exist to promote an attitude of cooperation between the community and local businesses. Where such efforts are extended to local consumers, a sense of mutual participation in the community's overall economic vitality is often achieved.

Professors Richard Wilkinson and Kate Pickett, from the Equality Trust have presented a lecture, with compelling evidence that among industrialized nations, more equal societies almost always do better in terms of health, well-being and social cohesion. Where large income inequalities within societies exist, that destroys the social fabric and quality of life for everyone, in the long run[21].

As Professors Wilkinson and Pickett studied data for industrialized nations, they noticed a clear tendency throughout the world for countries with large income inequality to score poorly on certain outcomes. They looked at a wide range of health and social problems and found that, outcomes are substantially worse in countries with the most inequality. Whether their findings are tested among the 50 states of the US, or internationally among the rich countries, there is almost always the same tendency for outcomes to be worse in more unequal societies. Their findings include the following:

Health and social problems are worse in countries with more inequality:

- Child well-being is better in more equal rich countries
- Levels of trust are higher in more equal rich countries
- The prevalence of mental illness is higher in more unequal rich countries
- Drug use is more common in more unequal countries
- Life expectancy is longer in more equal rich countries
- Infant mortality rates are higher in more unequal countries
- More adults are obese in more unequal rich countries
- Educational scores are higher in more equal rich countries
- Teenage birth rates are higher in more unequal rich countries
- Homicide rates are higher in more unequal rich countries
- Children experience more conflict in more unequal societies
- Rates of imprisonment are higher in more unequal societies
- Social mobility is higher in more equal rich countries
- More equal societies are more innovative
- More equal countries rank better on recycling[22].

We will not examine all these findings, but the discussion of a few should be developed further.

Income Distribution and Stewardship:

Why is the economy and income distribution so important to Earth Stewardship? The raw materials are resources. It takes the treatment of these raw materials to enhances and add value to the raw materials, as they are transformed into finished products. Many of those, whose jobs contribute most to that "value-added" component of the economy are compensated the least for their valuable contribution.

Most of the source of poverty comes from unpaid or underpaid wages. Throughout the world, poor people's work earn far more than what they are paid by their respective employer[23].

The withheld income, which creates the poverty, fails to feed the economy with the necessities, also deprives the economy of much needed spending power. Furthermore, the poor are deprived the opportunity to consume essential goods and services vital for economic growth. This absence of earned income continues to be a drain on the economy.

In a truly-free market economic activity would flow freely to engage and allow the participation of all potential partners in a fluid exchange. Too much money in the hands of the rich is not good for the economy. Large amounts of their money resources do not get into the parts of the economy that promote growth. Most of the excess money has been derived from the failure to pay fair compensation to those who actually add value to the economy.

Income Inequity Stifles Economic Growth:

Nobel Prize-winning Economist and Columbia University Professor Joseph Stiglitz is quoted as saying: "Our middle class is too weak to support the consumer spending that has historically driven our economic growth." The size of the US middle-class has declined significantly since 2007. In addition to those who were victimized by the housing crisis, many lost their jobs, due to the crash and were unable to find alternative employment in the shrinking economy[24].

In the US, the richest 400 people have almost half the country's income. Consider the implications of this fact. The majority 150 million people would spend most of their money in the economy, in ways that strengthen it and create jobs if that money were available to them. Most of the wealth of the richest 400 is invested abroad, or otherwise held out of the active economy. As such, those 150 million provide by far more value to the economy than the 400 richest.

In effect, roughly half the value of the US economy has little impact on the active, circulating, immediately accessible working assets that feed the real day-to-day activities, which support our daily lives. Accordingly, these 400 families are receiving more benefits and privileges from the economy.

Income Inequity is Harmful to other Aspects of our Wellbeing: In a functioning democracy, the people have significant control. The domination system is expert at subduing the democratic function. In the same way that modern civilization "controls nature" by running interference, the domination system is inclined to control by blocking progress. Their wealth becomes their weapon.

It should be remembered that the Wall Street crash of 2007 was, in no small part the result of a steadily growing income gap[25]. That pace of growth was further increased by the abuse of the "bailout" funds given to the large nationwide banks in 2009 to boost local economies.

The Urgent Need for a Genuinely Equitable Tax System:

In civilized society, people seek government to protect basic human dignity, ethical standards, genuine fairness and justice for all. In fact, the Declaration of Independence calls for such standards. However, in the spirit of a domination system, the authors of the declaration had a limited concept of "all men". Slaves, indentured servants, natives, women and children were not expected to be included in that declaration.

Figure 1.

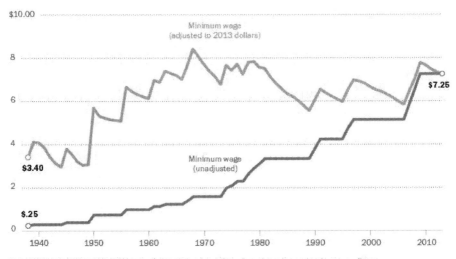

Federal Minimum Wage, 1938-2013

Shown in nominal (not adjusted for inflation) dollars and 2013 (inflation-adjusted) dollars

Note: Wage rates adjusted for inflation using implicit price deflator for personal consumption expenditures.
Sources: Bureau of Labor Statistics, Bureau of Economic Analysis, Pew Research Center analysis
PEW RESEARCH CENTER

We have come a long way from the exclusive mentality of the 18th century. However, the male dominant tendencies are not easily shed, especially among those, for whom power is everything. Yet, for those who wish for sustainability, it is necessary to uphold the Earth-centered way of life, which acknowledges the natural cooperative interdependence, which makes life possible and sustainable.

Earth stewardship calls for harmonious cooperation and collaboration. There are sufficient resources on Earth to be shared through an open economic system, which engages everyone to participate. Open communication is vital to feeding the process in harmony. Above all, honesty, sincerity, mutual respect and a will to serve each other enables a world to support itself through community. We do not live that way at present. We have to create forums that bring people together with the mutual desire to serve each other.

Professor Stiglitz recommends that tax reform would be an effective approach to making the economy reflect our democratic consensus. He suggests seven steps to promote growth and equity:

1. Raise Corporate Income Tax Rates while providing incentives for investments and job creation in the US
2. Reduce Spending on Corporate Welfare
3. Tax the financial sector
4. Tax on monopolies and other rent-based enterprises
5. Ensure that multinationals pay their fair share of taxes and have incentives to invest in America
6. Increase taxes on industries that produce negative externalities
7. Make dividend payments tax deductible, but impose a with-holding tax[26].

Reclaiming the Middle Class: The domination system is particularly wary of a middle class. Initially, it welcomes them as a convenient buffer from the poor. The system needs an educated clerical staff to manage the growing complexity of governance, as the system expands in its prosperity. Their service to the upper class is more valued than that of the poor and their income reflects the distinction.

The poor work traditionally in menial jobs, whose tasks are considered less valuable, although they are more basic. In times of economic stress, they become relatively more valuable, as their pay is low. Thus, more of the employer's profit will come from the work of his cheap labor. In such times, the laid off middle class sometimes drifts into poverty, sometimes supervising these low-wage workers.

Given the continuous campaigns by unions, workers, politicians and others to raise the federal minimum wage, it begs the question: Who are these minimum-wage workers, anyway? What are their wages? (See Table 1 above).

The Proposed Green Economy:

In 1970, when Wendell Berry published a collection of essays entitled: A Continuous Harmony, he quoted from The Triumph of the Trees, (by John Stewart Collis), who identified a historical trend in the evolution of human society in which we went from an Era of Mythology, through an Era of Economics, bringing us to the threshold of new Era of Ecology[27]. The green revolution is a significant beginning step toward this proposed Era of Ecology.

Since the rise in awareness of the accelerated global climate change, the coal and petroleum industries have flooded the media with misinformation, which has misled the public. In the meanwhile, most recent evidence points to the realization that predictions of climate disasters are more dire than previously anticipated.

Beware of the Corporate Pretenders: One movement behind the shift in economic emphasis is loosely referred to as the "Green Revolution". Persuaded by the argument that we could just tweak the incentives and penalties "a bit here and there with green taxes", and a new phase of corporate green enterprises would compete successfully.

Some environmentally enhancing innovations do contribute to net reductions in excess carbon build-up. However, many have been disingenuous. Any ones that propose a mere reduction in the rate of carbon build-up are ludicrous. We simply do not need the gasoline engine.

Nuclear energy is the most dangerous of all the proposed alternatives. It is too great a cost and environmental risk, whose only service is geared to offset a bad investment! The long term risk is not fully known and what is partially known indicates potential disasters, the likes of which we could not afford, given today's technological capacity. Our geological knowledge tells us that nothing buried is safe. The geology of diamond mines demonstrates this reality quite clearly[28]

By the end of the first decade of the 21st century, it was evident that the voluntary approach of the mitigation of CO2 and other greenhouse gases was a failure. An alternative approach was offered by many economists and environmentalists[29].

Furthermore, the greenhouse gas cap-and-trade schemes failed at great cost to poor farmers in many agricultural countries. In Mexico, for instance, farmers in one region were bitter and resolved to discontinue schemes, when they found that the money they were being paid to forsake their normal crop of maize was not buying the food they needed for their families. Subsistence farmers often function outside the money economy. They expected the money offered to go much further. In addition, since they did not grow corn, there was a shortage and the new price was even higher[30].

The green capitalism enthusiasts vastly underestimate the gravity, scope and speed of the global ecological collapse we face and thus unrealistically expected unlimited growth. With the introduction of the cap-and-trade plan a lobbying free-for-all was unleashed, which led politicians to dole out favors to various industries, undermining the environmental goals. After four years, it had become clear that the system had produced little noticeable benefit to the climate - but generated a multibillion-dollar windfall for some of the biggest polluters[31].

Against this backdrop yet another battle has been brewing. An organized coalition of conservative groups is pressing states to "fiercely resist" the Environmental Protection Agency's carbon pollution rule for existing power plants. In a war of words, they are out to block any legislation that limits their operations[32].

The rash of recent coal ash disasters, (see Chapter 20), leaves vivid examples of the price victim communities inherit in the wake of promises offered by the above coalition. In addition, several fracking incidents the petroleum industry would hope we forget are forcing us to trade scarce water for oil.

In the final analysis, our erroneous way of thinking engendered by the Doctrine of Discovery encourages us to rely on short-term political solutions to long-term, chronic problems. This quirk in our way of thinking is well documented in Chapter 21, below, on the Crumbling infrastructure

CHAPTER 21

————— ❖ ❖ ❖ —————

The Crumbling Infrastructure[33]

THE INFRASTRUCTURE REFERS to the materials, appropriate construction, systems and other conveyers of service critical to the survival of urban civilization. For instance, in order for us to be able to turn on a tap and receive water, a reservoir has to be connected to a system of pipes, valves, etc., which transport the water from the reservoir to my house. The system, which transports the water is infrastructure. Roads are essential to the operations of a city. The road system is infrastructure.

Urban society as we know it today could not exist and be sustained without a network of interdependent systems of infrastructure. The water system in most urban areas is heavily dependent on electricity to power pumps critical to its overall distribution system. Roads are an essential part of our transportation system. Indeed, our shopping centers would not be so well stocked with the incredible variety of goods and services we so easily take for granted, without the complex, diversified infrastructure network to serve them. We use those same roads to commute to work, school, recreation, etc.

In similar and somewhat surprising ways, the urban attitude of taking things for granted is at the heart of the current infrastructure dilemma. The infrastructure of the US is crumbling and its urgent repair is critical. Let us outline the nature of the dilemma.

The Infrastructure:

The US have the most widespread network of paved roads on the planet. Similarly, we have harnessed the power of flowing water to produce hydro-electric power on large rivers. Petroleum is the main power source used today to drive automobiles, as well as produce electricity, creating in the US the largest electrical power grid on Earth. The coal, which brought us out of the nineteenth century continues to be a major energy source. Yet, we live without the sense of dependency inherent in our actual lifestyle!

Such achievements make possible our convenient lifestyle. They contribute significantly to advances in public health, sanitation, electricity distribution, extensive highway networks, revolutionized airline transportation, computer systems, satellite traffic, improved education and skills training, plus a variety of advances in technology that complement the above benefits.

Many millions are employed in industrial societies to maintain such services. And, it is generally assumed that adequate management systems, regulations and proper maintenance are necessary to keep these services in continuous operation. The business community certainly depends on the infrastructure and takes it for granted.

Most of the US infrastructure was built during the first few decades after World War II. However, routine maintenance has not been applied consistently in recent years. This is because our leaders often compromise the appropriate funding of necessary maintenance. Today, many of these major systems are approaching the limits of their normal life expectancy. Accordingly, many systems are breaking down, normal repair and replacement are overdue, and appropriate correction would continue to be more expensive.

Our Modern Lifestyle is Not Conducive to Stewardship:

The conveniences of the modern global community encourages people to take them for granted. The resulting sense of "entitlement" has postponed an increasing likelihood of major catastrophes occurring in many large metropolitan areas throughout the world. Cities rely heavily on electricity, integrated sewer systems, grid-patterned streets, paved roads with curb and gutter amenities, houses, apartment buildings, modular homes, tap water, etc. that are integral to a coordinated network of infrastructural systems.

Routine maintenance requires strong stewardship instincts. Entitlement instincts encourage a valued sense of freedom without responsibility. This is the promise of an affluent society. Widespread infrastructural neglect is a natural produce of the spirit of affluence.

In contrast, the spirit of stewardship, continually prompts the natural question: "How do we use these resources so that they would be available forever?" Earth's lifestyle is based on balance. Authentic science is also based on balance: "Matter cannot be created or destroyed!" That balance is provided by a QUEST FOR SUSTAINABILITY.

Describing the Crumbling Infrastructure:

Stephen Flynn: author: The Edge of Disaster explains that we have been negligent and, in many ways are repeating the Roman experience. They neglected their infrastructure and paved the way to the Dark Ages[34].

The government offers to "balance the budget" by cutting many domestic programs, including infrastructure maintenance. Their concept of a "Balanced Budget' at a national level does not seem to take into consideration a pivotal fact. If our budget for the future excludes significant parts of our population and critical maintenance of our infrastructure, it is simply **NOT BALANCED**! Our road system is probably the most representative metaphor for future dilemmas, where short term decision-making clashes with long term responsibility.

What we built from the 1930s to the 1960s is truly remarkable. . . . . . Much of that construction is approaching the end of its life cycle. However, we have not been as diligent in maintaining as in building these masterpieces.

When interviewed in a documentary, Edward Rendell, then governor of Pennsylvania stated: "When President Eisenhower left Washington in

1961, 12.5% of US budget was spent on infrastructure. By 2008 we were spending about 2½% on infrastructure development and maintenance. China spends 9%. India spends 8%."

The U.S. Society of Civil Engineers graded the country recently on the state of its infrastructure: Bridges – C; Dams – D; Drinking water – D-; Levees – D-; Roads – D-; Wastewater D-; the Grid D+; Schools – D; Solid Waste C+;Transit D; Inland Waterways D-. The overall grade was a D. The Society estimated at the time that it would take $2.2 trillion to get our infrastructure back to respectable, safe levels. The 2009 budget had allocated only $72 billion for infrastructural repair in the Federal Stimulus package, according to Andrew Herrmann, Society spokesman[35].

Roadway Bridges:

Dan McNichol, in his book, The Roads that Built America, suggested that the country was headed in the same direction as the Roman Empire, which neglected its infrastructure and saw the Empire collapse as a result. The Roman road system enabled them to expand into a large area, which they could not maintain. Their viaduct system was the envy of their world. Their infrastructure enabled Ephesus to become the "New York" of its age[36].

Most modern roads and bridges in the US were built 50-70 yrs. ago, with an average life expectancy of about 50 years. The average age of existing infrastructural components in the US in 2014 was about 48 years! Many bridges were not designed built to handle today's traffic volume or intensity. Bridges are especially neglected. In cold climates, the salt used to melt snow can corrode the steel in the structure.

The collapse of the I-35 Bridge in Minneapolis on August 2007 was a wake-up call! Built in 1967 a critical flaw was overlook during inspection of the bridge while under construction. A gusset plate, which holds together the steel beams supporting the bridge structure was a half inch thinner than what the plans called for. Inspectors overlooked it, dooming the structure.

The Tappen Zee Bridge spans the Hudson River 13 miles north of New York. It was built in 1954, during the Korean War, when much of the steel needed for its construction was not available. Accordingly, many of the supports are wooden. The structure was designed for a 50-year life. Wood poles can last as long as concrete and steel, but they have to be treated and buried in soil low in oxygen content – Tappen Zee poles were not treated. Also the bridge was not designed with basic drainage, allowing water to pool in places and increase corrosion.

Update on the Tappen Zee Bridge Status: After the initial proposal to replace the old bridge, (1999), a new proposal to replace the Tappan Zee is underway. The first span of the new twin-span bridge is expected to be opened in 2016, and the second span should be complete in 2018. "The new bridge, The New York Bridge, will be designed and constructed to last 100 years without major structural maintenance[37]. The proposal is quite impressive and in time to relieve the present situation on the existing Tappen Zee, which was designed to handle 100,000 trips per day. The traffic in 2008 was at 138,000 trips per day.

Levees:

The 1993 Mississippi flood broke through dozens of miles of levees. Hurricane Katrina exposed more critical under-maintained levees along the Mississippi. Nationwide, 100,000 miles of levees protect millions of acres of farm, towns, cities and other habitat. The levee system in the US represents one of the greatest potential disasters in the country.

The Sacramento levees are a particular problem, because they were built haphazardly by farmers to protect croplands in the Sacramento Delta. Delta land is very fertile. However, it is also generally waterlogged. As the water is pumped out of the soil, it shrinks. So, dikes are built to keep the water out as the ground sinks.

The continuous draining of the delta soils over the past several decades has shrunk the soil, leaving most of the farm land 12 to 25 feet below the normal flow of the surrounding rivers. The delta levees are so weak in many areas that, if they were not being drained by pumps working around the clock, the land behind the levees would fill up with water on their own, in three months[38].

The Army Corps of Engineers, (ACE), did not build them and their structure is inadequate. They have been leaking and many have failed in recent decades. The region is most vulnerable to hazardous infrastructure failures in North America.

Kevin Knuuti, U.S. Army Corps of Engineers, said in an interview about levee protection that some 177 levees in several states are at risk of failure. Jeffery Mount, a geologist at the University of California, Davis, who has studied the condition of levees in the Sacramento region, accounts for almost 200 miles of levees, protecting well over 100,000 people. Some subdivisions could find themselves under 20 feet of water in the event of a complete failure of critical levees.

The massive floods of 1980s and 90s exposed weaknesses in the Sacramento levee system. These levees were mostly built by farmers in the early 1900s, using whatever was available to them locally. Usually, their construction was made up of a mixture of sand and gravel sufficient to protect crops. Many of these levees were breached during the floods of the 1980s and 1990s.

The ACE has been helping to shore up the Sacramento levee system in recent years. They use a mixture of sand, silt and clay, adapted to be effective for the particular soil types of each given location. Clay by itself would provide the most impervious barrier. However, the levees need to be able to leak slightly to relieve the internal pressure that would normally build up in such structures. The Corps project used heavy rollers to pack down the levee construction, which will make it far sturdier than that done by the farmers.

Nevertheless, this project is not enough, and as of June 26, 2014, the levee reclamation and improvement program had not yet begun, as the USACE and the State negotiated funding responsibilities for various component parts of the project. By all estimates, this is expected to be a multi-year project, when it eventually gets underway.

There is no official inventory of the country's levees, which is being compiled since Hurricane Katrina in 2005. That project will not be completed for some time. Thus, in the context of homeland security, we do not know how many levees exist, where they are, in what condition and/or how many people are at risk. As discussed earlier, (see: Chapter 11), floods are the most casualty-prone "natural disaster".

Levees are somewhat like dams in their vulnerability to the bodies of water they attempt to hold back. If the levees are porous, they can eventually fail. The California Delta is formed at the confluence of the Sacramento River, the American River and the San Joaquin River, as they flow together into the Suisun Bay in Central California. This provides a huge volume of water flowing into the San Francisco Bay, via the Carquinez Strait.

The Delta has been heavily farmed for a few hundred years. Being a delta, fed by considerable sediment, deposited by the above three rivers, its lands are extremely fertile. During the extensive dry seasons of Central California, the lands are well above the flow of the surrounding rivers and available for farming. The region receives almost all its rain in the winter. Most of it is stored in the Sierra Nevada as snow, which melts in the spring, occasionally sending a generous flow of water through the delta and flooding it. Early farmers built levees to protect their crops from being flooded.

Since the Gold Rush, when farming activities increased dramatically to feed the rapidly growing city of San Francisco, the farms have been protected by a series of levees. The City of Sacramento also grew during the Gold Rush of the 1840s and 1850s and some of the farm land has been taken over by the growing metropolis. In recent decades, some of this farmland has been converted into suburban residential subdivisions. Natomas Township is one such suburban settlement.

Protected only by the current low-grade levees, Natomas is home to 70,000 people, near the Sacramento Airport. Along with many other areas not included in the ACE project, this area would be flooded by a major levee-failure event.

Doug Thompson, California Department of Water Resources, says that a 440-mile series of aqueducts serve most of the farmland in the southern Central Valley with fresh water from reservoirs, fed by Sierra snow-melt. If there were a major breach of the delta levees, (a possible scenario resulting from a major earthquake in the immediate surroundings), "the suction effect of the dissolving of the deltas would be so great that it would pull 300 million gallons of salt water from the San Francisco Bay into the Delta".

That salt-water would contaminate up to half of the state's drinking water. It would take 3½ years to replace a water supply that currently serves 25 million people. This is not the part of the California levee problem that the Army Corps of Engineers is working on. To build safe levees throughout the Delta would cost between $25 million and $40 million per square mile. The state has no plans to adequately "quake-proof" the faulty delta system.

The gates at Clifton Reservoir, the main reservoir serving the aqueduct), would be closed to protect its fresh water. The quake impact would, however, stop any further supplies of fresh water to the reservoir, which would be rationed and would last only six to 12 months. It would take two to three years to replace that water supply system.

The farmers are the primary source of ongoing inspection of the levees. California has no system in place to protect against such an earthquake. Accordingly, further farming in the San Joaquin Valley, one of the nation's largest fruit suppliers, would cease.

Dams:
Many of the problems facing the levee systems are identical with those challenging some dams. Like levees, dams have often been constructed

hastily. As such, they are often not equipped to handle flood conditions on the river on which they are located.

Dams serve a variety of functions. The most common use is for water storage, (for a reservoir, often as a limited-use lake). In this setting, the dam can also be used for flood-control. Sometimes, the dam may be an integral part of a hydro-electric power plant, as well. Specialized dams are used in some cases to keep hazardous materials from flowing into a natural water-course. In coal-mining districts, coal-ash dams are constructed to hold coal sludge from flowing into nearby rivers.

The Status of Basic Dam Maintenance: There are 85,000 dams in the US, 4,000 are unsafe and 1,800 of those are high-hazard dams. This number of unsafe dams has quadrupled since 2001, as the actual status of more of these dams is being documented. Dams are on average 51 years old, close to the design life of many of the smaller ones. Regular inspections are critical. However, qualified inspectors are often available. In 2007, Texas had 7 inspectors for 7,400 dams. The inspectors examined 239 dams. That year, Iowa had one full-time and one part-time inspector, overseeing 3,344 dams, of which 128 had been inspected. Alabama had no one to inspect their 2000+ dams. In addition, orphan dams abandoned by private owners or corporations are not inspect.

Wolf Creek Dam: Its face is over a mile long. It is the largest dam east of the Mississippi. Built in the 1940s on porous limestone, it leaks so badly that a group of engineers thought it would fail in five years. It has been patched a few times since. The dam holds enough water to cover the State of New Jersey with 1 foot of water.

Leaks were found on the dam face and the level of the lake was lowered as a precaution, as they tried to fix it. Mike Zoccola, US ACE tried to downplay the danger. Five other dams in the Eastern US were also built on porous limestone in the 1930s. Wolf Creek Dam holds back 6M cu ft of water in Cumberland Lake, which would enter the Cumberland River in event of a dam failure. A wall of water would travel down the river at about 25 miles per hour. In 1968 large sinkholes appeared on the grassy face of the dam, 15ft in diameter and 40ft deep. People said they could hear water running in the sinkholes.

The ACE had pumped millions of gallons of grout into the holes to plug them up, but the seepage continued. In 1975 the ACE built a 2,000ft long barrier into a trench cut vertically into the dam face to create a more formidable interruption of the seepage. But in 2004 new wet spots appeared, from seepage around and under the wall on the dam face. They poured more grout into the critical areas around the wall and began to prep for a completely new wall in front of the existing barrier. This wall is being built deeper than the first and will go along the entire face of the dam. They claim that it is safe until 2012, when they expect to complete construction.

According to the estimates of Don Franklin, Kentucky Emergency Management, if a heavy rainfall soaked the area for days, it may soak into the limestone foundation of Wolf Creek Dam and create several sinkholes. They, in turn could accelerate underground water movement, weaken the dam surface, causing a breach!

Consider this scenario: A dam breach, about 600 ft wide and over 150ft deep could quickly be eroded by the force of the escaping water. That would send a 20 ft wall of water down along Cumberland River, taking everything in its path – a formidable flash flood. It would cover the 280 river miles to the City of Nashville in roughly 2½ days. The people of Nashville would have time to evacuate, but the impact would be enormous[39]!

The flood of 2010 in the Nashville area gave a warning and an illustration of what a Wolf Creek Dam break could do to the city. But, a dam break would be much worse, since it holds back much more than two full days of rain water. Probably in response to this warning, the Army Corps of Engineers under took a major project to build a 275-ft-deep, 3,800-ft-long concrete wall composed of piles and panels within the clay embankment and into the rock of the dam[40].

A similar smaller-scale fix was attempted in 1976. This time, the difference is in the barrier wall's greater mass and depth as well as the materials and methodology used. "The dam itself is in top condition," says Tommy Haskins, the Corps' geologist and technical manager. "They did a superior job [in 1976] on the embankment. If they hadn't done that, it would likely be gone. The problem here is in the limestone foundation and the depth and construction of the core, or cutoff trench. The cutoff trench was not only ineffective, it serves as a conduit for seepage"[41].

Thus, by his own assessment, this Army Corps manager is pointing out the primary reason why one does not build a dam into limestone bedrock. However, people are recognizing the potential disaster represented by this dam and are diligently preparing for the inevitable. An Emergency Response Team appreciates the magnitude of the potential disaster.

Cecil Stout, Director of Emergency Services for the Clarksville-Montgomery County Chapter of the American Red Cross, and logistics coordinator Mike Vogt are a part of the Emergency Response team. The Red Cross office and thousands of homes and businesses along the city's rivers could be underwater if the dam fails. They have already mapped out evacuation routes to strategic location in safe areas to accommodate dislocated residents along the river[42].

Wolf Creek is one of the five most vulnerable dams in the country. If Wolf Creek, with a high risk of failure, were to break, parts of Clarksville will be underwater in about 33 hours. In another 24 hours or so the wall of water would hit Nashville, inundating all river-front settlements along the way and beyond. The warning from the 2010 flooding event continues to be ignored.

Here is the fatal flaw in the domination system of our modern lifestyle. We make a wrong decision. We discover that it is wrong and why it is wrong. However, we have invested so much in it that we have extreme difficulty in admitting to the futility of "staying the course".

The dilemma is often further confused by those decision-makers, who have direct monetary benefits in "staying the course". Apparently, some level of political pressure comes from up-stream recreational activities. Voices continue to be raised by locals with the long term interest of the region in mind. In response to the Army Corps report on their latest upgrade of the dam, one reader called for the dam to be closed[43].

In addition, other concerns related to the initial building of the dam and its interruption of the natural valley loom large in the potential reclamation of the original river course. When one removes a dam from the river, the drowned valley upstream has to heal. Often, the greater upheaval comes when the river reclaims its valley downstream.

Everything built on the banks of the renovated river, for some miles downstream from the dam is flooded out. That is one problem, somewhat separate from another physical uncertainty: The silt accumulated behind the dam during the life of the dam is washed downstream to settle into an unstable new river, which, for many years would be resettling its banks[44].

Disposable Coal-ash Slurry Dams:

When we think of dams, we picture clean water reservoirs. In contrast, coal-ash dams hold back highly toxic waste from coal-fueled electric power plants. The coal-ash dam takes away the impurities, which leave behind a new fuel, better known in some alternative energy circles as "Clean Coal". One would not expect a coal-fueled power plant to require a dam in its operations, but without it, "clean coal" is not possible.

However, the process of cleaning the coal requires the removal of many of the ingredients that would make the use of coal a more polluting process. The coal used to power energy plants also produces a wide range of toxic byproducts, which have to be disposed of safely. Because coal-ash dams are extremely toxic, these dams present a major threat to the environment.

So far, this chapter has been observing the state of maintenance among major elements of our infrastructure. Coal ash dams are not, in themselves normal infrastructure. However, the challenges they pose are sometimes similar to the flash floods that may accompany a sudden dam failure.

Sue Sturgis prepared a report on May 20, 2011 for the Institute for Southern Studies, updating the event of 22nd Dec 2008, when a coal-ash dam run by the Tennessee Valley Authority in Kingston breached, spilling 1.1 billion gallons of toxic waste across a 300 acre stretch of the Emery River in eastern Tennessee. The area was evacuated. 42 homes were destroyed, but there were no casualties[45].

Coal-ash is the powdery residue left after coal has been used to generate power. It is stored in a surface impoundment, like the one that collapsed, in Kingston. The coal-ash sludge that was released, contained large quantities of toxic heavy metals, combustion byproducts and radioactive elements that washed into a neighborhood and two nearby rivers.

WADE PAYNE/AP: Homes destroyed by the TVA's
December coal ash spill in Tennessee

Since then, the EPA has identified and inspected 676 coal ash dams, nationwide. Of those inspected, 55 got a "poor" rating, which means "they need repair or further testing, or lack proper engineering documentation".

Of the poor-rated dams, "nine are deemed high-hazard, where a breach would likely kill people". In all, 39 were judged to pose a significant hazard from which a breach could "cause substantial economic and environmental harm".

These hazards are generally triggered by an excessive rain event that has created flood conditions. It is worse than a natural flash flood, since toxic material is mixed into the mud-flow created by the flood. Chapter 20 discussed in greater detail hazards inherent to mining and other risky processes involved in obtaining minerals on an industrial scale.

Underground Water Mains:

It is easy to forget the maintenance of underground utilities and hidden infrastructure. Underground pipes are at major risk throughout the country. 800,000 miles are literally bursting apart each year, because of overdue maintenance.

Even the leaks and explosions often escape the awareness of the general public. Occasionally the news media carries a story of a major break. However, the nature of these reports gives the impression that each event is an unusual, isolated incident. These events are not merely wasteful; they can be deadly!

One Tuesday in Maryland in 2008, a few days before Christmas, there was a break in a 66-inch pipe. It occurred on a suburban roadway, at a time when traffic was light. Had it occurred on a heavily travelled roadway during the rush-hour the loss of life could have been disastrous. The Montgomery County incident, (above), may have seemed like an isolated incident. But, the county had seen 4,000 breaks in two years; (over 477 breaks in one particular month)[46].

Most breaks occur during the winter, aided by the cold weather. In addition to the stress of freeze/thaw cycles, an 8" pipe, unattended for 20-40 years can build up internal rust, sufficient to completely choke off the flow of water in that pipe[47].

Old pipes, in "Rust Belt" cities east of the Mississippi are particularly vulnerable. Estimates put the annual water-main breaks at 240,000 nationwide, according to Sunil Sinha, Professor of Engineering, Virginia Technical Institute. The leaks are not merely a waste issue. As such, very few of these breaks can be considered "isolated events".

Contaminants can get into the water stream via the very leaking points. Tuberculation is common in many old systems. It is a process by which tubercles build up inside the pipes, trapping microbes, which can contaminate the water. Many cities have had to issue "boil-water" advisories, which are becoming common in some cities[48].

In 1993 400,000 people in Milwaukee, (25% of the population), got sick, when a harmful parasite from animal fecal matter got into the city's water supply. This was the largest water-born outbreak so far recorded in the US. At least 100 people died and schools had to be closed.

Leakage: Two fires raged out of control in Washington D.C. in 2007 due to fire hydrants that did not deliver water, because of unknown leaks. Inspectors found, upon further investigation, that about 20% of their fire hydrants were non-functional[49]!

In all, water pipes in the US leak 6 billion gallons per day, sufficient to adequately supply the state of California! New York City loses 10% of its water in leaks. Atlanta loses 14%; Buffalo loses 40%! Moreover, all these figures could increase exponentially, as more systems succumb to old age.

Consider where such volumes of water would go in a city. A case in point is the system, which feeds New York City: 1.2 billion gallons of drinking water/day, via a system of aqueducts. The Delaware Aqueduct is 85 miles long. It is one of the longest aqueducts in the world, 13.5 ft. in diameter. Currently, this aqueduct leaks between 10 and 36 million gallons/day into the ground around it. Whole communities are drowning in the leakage.

Wawarsing, N.Y. is such a community in a limestone area that happens to be just above a part of the aqueduct that is leaking generously. Accordingly, some 50 families are bombarded by a host of seepage problems. The result has been a host of flooded basements and occasional sinkholes. They have wells to provide their fresh water and septic tanks for sanitary sewer needs.

All these facilities have been overwhelmed by the aqueduct leakage. Water cannot be used for drinking! E. coli has been detected coming out of tap. During construction the section that was built under Wawarsing and ran through a limestone base leaked so badly that they had to line it with steel and seal it with concrete.

New York is not ready to face the eventual shut-down challenge. One such solution would be to abandon that entire system and build another that would not pass through limestone rock at any point in its construction. After years of study, a solution was proposed and is currently underway[50].

Storm and Sanitary Sewer Systems:

The US has over 1 million miles of sewer with an average age of 50 years, much of which is leaking. The biggest problem is that sewers are buried underground, which often makes it difficult to locate leaks. This problem can be solved by installing sensors in the pipes each time actual repair is undertaken, often cutting the overall cost of future maintenance.

Sewage is more destructive to pipes than drinking water. Raw sewage contains hydrogen sulfide and is easily converted to sulfuric acid, which eats away concrete and steel. Sanitary sewer pipes are not pressurized. Therefore, it is easier for them to be infiltrated by ground water from cracks and breaks caused by tree roots, etc. Decay can be further accelerated by excessive rain water runoff getting into the sewers. Moreover, root and other infiltration can block drainage, causing backups into people's homes and other living spaces[51].

St. Louis has excessive sewer challenges. The city is burdened by an outdated system of combined storm and sanitary sewer pipes, which get overwhelmed during a heavy rain episode. This often causes raw sewage to be dumped into rivers, streets, homes, businesses and anywhere connected to the sewers. The city dumps 13 billion gallons of combined storm and sanitary sewer into the Mississippi and other local streams, annually (enough to put the entire city one foot deep in waste)! Nationwide the problem is over 190 billion gallons dumped into public waterways, annually.

It is estimated that over 700 cities have antiquated combined sewer systems and most are plagued with sewer back-ups. Given the state of sewer maintenance nationwide, sewer back-ups are becoming for many a new way of life!

Many cities and towns are burdened by inadequate tax bases. Deferred maintenance is as much a habit of local government as of federal agencies. Deferred expenditure only causes the maintenance cost to increase with time. As such, local communities often find the cost to catch up with urgent maintenance has become prohibitive.

In many cases, their dilemma has been partly the result of federal inefficiencies. Such setbacks incurred by local projects may occur, when contract payments from the federal agency funding the project are late. This dilemma may be further frustrating to local communities, when delayed funding for one project causes delays in related projects, raising the actual cost of operations to exceed the projected cost.

St. Louis, like many other cities, has received federal fines for the pollution caused by deferred maintenance on the sewer system. This further increases the cost of maintenance, which is passed on to residents in service fees. Thus, in a twist of fate, the poorest communities, which are often the most frequently victimized by interrupted services, (for example, sewer back-ups), are further burdened by the rising service costs.

The Roads:

The road system upon which we have built our wealth and prosperity is literally falling apart. We have not kept up the required ongoing maintenance on them! A major collapse nearly happened in March 2008, on the I-95, a major highway, at a point just outside Philadelphia. A worker on a lunch break noticed a crack 2 inches wide on one of the elevated highway support columns. The highway was shut down immediately, possibly averting a calamity many times greater than the collapse of the Cypress Freeway in Oakland, CA on October 17, 1989, caused by the Loma Prieta Earthquake[52].

Two-thirds of the country's busiest roads are 40 years old, or older. The majority of these roads are designed to last fifty years. What is the status of the nation's highway maintenance budget? Michigan is so hard pressed that it is now allowing some secondary paved roads to return to gravel.

The aged infrastructure is handling increased demand. In some areas, highways are handling traffic loads in excess of what they were designed to handle. The deferred maintenance is so bad, that many state budgets have not sufficient funds to catch up with mere maintenance! Some workers refer to this type of maintenance as "Patch and Pray". And, some interstates are so bad that they have to be completely replaced, (for example, I-66 in St. Louis).

The US has paved 4 million miles of roadways. That is enough to circle the planet 160 times! 47,000 miles of them are interstate highways. Beginning in 1956, the work undertaken nationwide to build the freeway system was the largest earth-moving project the world had ever known. 1.6B tons of concrete, asphalt, sand and gravel went into the project, (3-times the weight of humanity!). This highway system has been the backbone of commerce, industry, daily commutes, recreational travel and any number of miscellaneous trips we depend on daily. Shouldn't we be maintaining these roads regularly?

Power Grid:

On August 14, 2003, it took a mere 12 seconds for a black out to occur, which kept 55 million people in US Northeast and parts of Eastern Canada without electric power for two days. On that day, a power line sagged in Ohio and fell into a tree that should have been trimmed. It shorted out the related electric grid, which, in turn shut down one grid after another, cascading adjacent grids, until six states and Ontario were blacked out. That means that much of the other infrastructure was interrupted on a sweltering hot day in the Northeast.

Grid operators can normally transfer power to alternating routes, when one goes down. A software glitch in one of the Ohio grid networks had slowed down, so that the switching fell behind causing major circuits, one after the other to shut down. Such an event automatically shuts down many other forms of infrastructure throughout the affected regions. For instance, subway trains stopped dead in their tracks. Sewage treatment plants, the pumps that keep the subways from flooding with ground water, home electric health units, which keep patients alive, etc. are powered by electricity.

The US has about 10,000 power plants and 160,000 miles of power lines. However, the grid is overwhelmed by aged technology. Some systems are overloaded, while others are slowed down by dated equipment, sometimes at the point of inevitably breaking down. Many parts of the grid have not caught up with the digital age. Loads have increased faster than the overall effort to maintain them adequately, or further expand services to keep pace with growth. Even basic things like power poles and transformers are literally decades beyond their replacement schedule.

Most of the 160M wooden poles used in the country are made from southern yellow pine. The poles are treated with an insect repellant. But the cold weather of northern states is hard on the poles, ice loads and freezing rain can add to stress on the lines. Tree limbs snap and fall on lines which often causes the poles to tilt. As poles tilt, power lines sag into trees and short out, causing local blackouts, which have doubled nationwide since the early 1990s. Demand for power has spiked 15% in the last decade, at a time when maintenance, replacement and newline construction are lagging behind sustainable schedules.

The average blackout experienced each year in the US lasts 214 minutes, compared with 70 minutes in Britain and 6 minutes in Japan, (reported in 2008). Furthermore, a stressed-out system can create local nightmares. Consider Deerfield IL, an affluent suburb of Chicago with a population of 18,000. They have experienced nearly 1400 blackouts since 2000, (average 200/yr.). These blackouts lasted from a few minutes to a couple of days. This is the impact of maintenance and necessary repair in systems that are not keeping pace with the rising demand for equipment conformity.

Technology exists to monitor the grid more closely for situations that may develop to interrupt service. Such a system can identify impending failures and address the problems ahead of time at much lower a cost, significantly reducing real maintenance cost.

However, the biggest, most extensive grid in the world is being held back by critical older parts of the grid, which are outdated and incompatible with modern parts of the grid in the digital age. The entire system needs to be more uniform and compatible with the most advanced technology. It could cost $1.5 trillion nationwide to make vital improvements to the grid, or the Deerfield experience might soon become commonplace through much of the country.

Other countries have completely updating their systems. They have made the commitment and invested smaller, but sufficiently updated systems to keep up with a higher technological capacity for their respective communities.

Technology now exists, capable of ongoing self-monitoring of entire systems. Such systems can identify impending failures and address the problems ahead of time at much lower a cost, increasing real maintenance efficiency significantly. "Prevention is still better and cheaper than cure!"

Our culture of indefinitely deferred maintenance is taking us to the brink of numerous catastrophic infrastructural meltdowns. We need to have a uniform electrical infrastructure with integrated connections, which would allow for a much more efficient national system. The cost of not taking these steps is prohibitive!

According to the American society of Civil Engineers, it would take $3.6Trillion by 2020 to raise the country's grade point average from a "D" to a "B". That is just to catch up to functional sustainability. And that does not take into consideration the potential impact of global warming[53].

Maintenance is the Lifeblood of Infrastructure:

Any infrastructure requires a schedule of routine maintenance, in order to continue to perform in the way it was intended. In other words, an infrastructure that is not being routinely maintained, is probably declining. Its core is not being sustained. It is on the way to breaking down.

Grade Sheet: Infrastructure Investment Needs presented by the American Society of Civil Engineers, 2013 [in billions of Dollars]

Infrastructure Systems	Total Needs	Estimated Funding	Funding Gap
Surface Transport	$1,723	$877	$846
Water/Waste water	126	42	84
Electricity	736	629	107
Airports	134	95	39
Inland Waterways & Ports	30	14	16
Dams	21	6	15
Hazardous + Solid Waste	56	10	46
Levees	80	8	72
Public Parks & Recreation	238	134	104
Rail	100	89	11
Schools	391	120	271
Totals	$3,635	2,024	1,611
Yearly Investment Needs Billions of Dollars	$454	253	201

At some level, an infrastructure, such as an electric grid, a highway network, a sewer system, or a wireless network, is built on a self-sustaining mechanism, which fuels and maintains its own routine operations. One turns a switch and a light goes on. You turn on a tap and water flows. The light goes on and the water flows because the infrastructure that was build and designed to deliver the service of water, or electricity is working.

Some systems, by their very nature, require more immediate routine maintenance than others. Some are more complex than others. For instance, a road network can continue to provide service without routine maintenance for years. Its quality will deteriorate, but it may still be useable. A storm sewer system can function, especially in arid rural areas for decades, with little immediate routine maintenance. In contrast, the electric grid requires routine maintenance every minute of the day.

Living On Borrowed Time:

Our infrastructure maintenance, (which is also our infrastructure management), is very far in debt. Future planning will not only have to catch up with current maintenance neglect; we also have to "pay off the loan" of time and resources implied by past neglect. What would have been corrected with minor repair and routine capital re-investment ten years ago, may need to be completely rebuilt today.

Moreover, we need to adapt to the growing impacts of global warming at its accelerating pace. In many cases, this requires a quantum leap beyond our present management standards. That is the next frontier in our progressive expansion.

New corrective measures will have to be added, such as sea walls and building retrofits. Sensor technology already exists to alert us to the need for routine maintenance adjusted to a calculated schedules. Rooftop gardens "reduce the carbon footprint" and green space can be designed to reduce water runoff and recycle storm water, especially in urban areas.

Photo by Michael Waldrep, May 14, 2012. *This article is excerpted from the SPUR report* **Public Harvest.** *Read the complete report at* spur.org/publicharvest >>

In terms of infrastructure costs, we can engage communities and neighborhoods in numerous civic projects, such as the Acorn tree-planting activity of Oakland, California in the 1980s. Many inner-city neighborhoods are teaming up with county agencies to learn composting and other efficient farming practices. Cooperatives and other mutual support groups are learning to grow, or improve community gardens growing organic food[54].

Notes

1. In 1996, when the federal minimum wage was raised to $7.25 per hour, Herman Cain, National Restaurant Association board chairman "successfully pressured lawmakers to have the minimum wage for tipped employees separated from the increase and kept at $2.13": This was a part of a report entitled: "Minimum Wage for Restaurant Servers Remains Stagnant for 20 years Under Industry Lobbying", by Dave Jamieson for the Huffington Post, (06/02/2012). The NRA further assured Congress that: "most servers already earn well above the federal minimum wage of $7.25, and that raising the tipped minimum wage could hurt kitchen employees and others who don't work for tips".

2. To be realistic and to avoid continuing to create poverty, a minimum wage has to be set as a living wage. Furthermore, it must be set to adjust annually to the current cost-of-living. It is futile to quote a monetary figure. An intentional minimum wage would have to be established as a living wage tied to the cost-of-living index. Consider, further that the law to suppress the wages of restaurant workers was passed in 1996 and holds almost 20 years later. Did the promoters of this law suggest, or promote a suggestion to the public to increase their tips to keep pace with the increase in the cost of living since 1996?

3. The contrast is vivid. Comparing wages paid by McDonald's in three countries it becomes abundantly clear that the fast-food giant can afford to pay a living wage. In the Philippines, where the workers are not unionized, the pay is $1.32 per hour. In Denmark, where the Government has set the minimum wage at the equivalent of $21US per hour, in January 2014, a BigMac cost $5.18 in Denmark compared with $4.62 in the USA. In New Zealand, they pay the minimum wage equivalent of $12.35 per hour. See: "Working At McDonald's Is Starkly Different In These 3 Countries", by Kevin Short, The Huffington Post, 15 May 2014

4. Cornel West interviews several prominent contemporary figures in an effort to instill a sense of hope in the minds and hearts of a discouraged generation: Cornel West: Restoring Hope, Conversations on the Future of Black America, Beacon Press, Boston, Massachusetts, c1997

5. The Parliamentary Reform Act of 1832 gave voting rights to male householders with land worth over £10 per year. In the boroughs, this included tenants as well as landowners. In the towns, it applied to owners only. Six out of seven adult men were left with no voting rights. Women were specifically excluded altogether. Excerpt from Socialism Today: "The Fight for Universal Suffrage",

Issue 120, July-Aug. 2008, published by Women's Social and Political Union.

6. Anup Shah has prepared a study on global poverty for the United Nations, which looks at the phenomenon from a variety of vantage points. See <u>Poverty Around The World</u> — Global Issues by Anup Shah. Last Updated Saturday, November 12, 2011: http://www.globalissues.org/article/4/poverty-around-the-world.

7. Banks and major corporations are taking taxpayers money overseas, to invest and in some cases paying wages overseas that are higher than what these same employers pay here for comparable work. In other cases, the same company pay workers in the poorer countries far less than the depressed wages paid here. This phenomenon is well documented in the report: "Big Banks Don't Pay A Third Of Tellers Enough To Live On": Study by Jillian Berman, The Hiffington Post. 23 January 2014. Also: "Working At McDonald's Is Starkly Different In These 3 Countries", by Kevin Short, The Huffington Post, 15 May 2014

8. Table 1 shows an estimate of the number of minimum-wage workers, by occupation in 2013. Compiled in a Pew Center publication: "Who makes minimum wage?" By Drew DeSilver, Pew Research Center, Sept. 8, 2014.

9. Figure 1 illustrates how middle-wage losses translated into low-wage gains, when many of the laid off middle-wage workers eventually found jobs. See: "Low Wage Job Growth": Data taken from: National Employment Law Project, August 2012 Report

10. The story of Chattanooga's recovery from its major pollution days is contained in: "Cinderella Story: Chattanooga Transformed", by Daniel Glick, National Wildlife Magazine Feb/Mar 1996

11. Originally designed as an outlet for college architecture students to practice in a traditional urban environment, the studio grew into a quasi-governmental project, with support from the city, the University of Tennessee, the Lyndhurst Foundation and the Chattanooga-Hamilton County Regional Planning Agency. Excerpts from Elis Smith's report on Chattanooga's Renaissance: "Stroud Watson: The Man Behind Chattanooga's Downtown Revival", by Ellis Smith, Chattanooga Times Free Press in January 2010, esmith@timesfreepress.com

12. Downtown, which had been a virtual ghost town after regular business hours, became more of 24-hour shopping, dining, and business venue. Downtown Portland, from Wikipedia, the free encyclopedia, last modified on 25 July 2014

13. The improved infrastructure included new water and sewer pipes. The streetscape and facade improvement programs, construction of Town Plaza, dredging, restoration of Suisun Channel and wetlands in the adjacent Suisun Marsh completed the plan execution. The details are described in: "Origins and History of Suisun City's Downtown Redevelopment," by Simmons B. Buntin, founder and editor-in-chief of Terrian.org: A Journal of the Built & Natural Environments, Issue No. 3. © Copyright 1997-2013

14. The Doctrine of Discovery kicked off the "official" European Colonial Era, in which Pope Nicholas V, as the Papal Head of the Holy Roman Empire, called on the monarchs of Europe to attack the people of the Americas and Africa as "enemies of God", because they were not Christian, (by the Pope's definition) The Doctrine of Discovery is a legal principle which has passed from Church law into common law. It granted the European discovering powers, a kind of unrestricted sovereignty over the lands and peoples they "discovered". It began with a series of papal bulls issued by Pope Nicholas

V in 1452 and Pope Alexander VI just after Christopher Columbus returned from his first voyage.

15. The World Council of Churches, The Episcopal Church of the USA and the Quakers are among many Christian groups which have formally renounced the Doctrine of Discovery. Nevertheless, many of the currently used U.S. laws derived from the Doctrine are still in use. The Quaker Indian Committee expressed its support for the U.N. Declaration on the Rights of Indigenous Peoples, which was adopted by the General Assembly Sept. 13, 2007. The Declaration presents indigenous rights within a framework of human rights. Only the U.S., Canada, New Zealand and Australia – countries with large populations of indigenous peoples with huge aboriginal land claims – voted against the Declaration's adoption. Australia has since adopted it. See the following reports: The Doctrine of Discovery, Manifest Destiny and American Exceptionalism, by Christian Hegemony, posted July 21, 2015. Quaker Indian Committee disavows Doctrine of Discovery, affirms Declaration: By Gale Courey Toensing, updated: Dec 17, 2009. Doctrine of Discovery, Used for Centuries to Justify Seizure of Indigenous Lands, Subjugate Peoples, Must be Repudiated by United Nations, Permanent Forum Told, 8 May 2012

16. His mission was to "spread of the Gospel by peaceful means alone". He challenged the church to understand true Christian doctrine and to respect and utilize native cultures as part of the missionary enterprise. See: Bartolome de las Casas: A Brief Outline of his Life and Labor, by David Orique, O.P.

17. This document proved a powerful weapon in the hands of the pro-Indian forces, although it was never formally published in the Spanish dominions. The first step to repudiate the Doctrine. It is described in: Doctrine of discovery first repudiated in 1537, by Michael Swan, The Catholic Register, October 2, 2014.

18. The earliest accounts of the Pilgrims settling in Massachusetts carry graphic stories of epidemics. In one written by a colonist, John Winthrop commented in 1684 that most of the natives in their region died of small pox between 1618 and 1619. See the following publications: Epic World History series: <u>Epidemics in the Americas.</u> Also: <u>Indian Nations of North America</u> by Rick Hill, Teri Frazier, George Horsecapture, for National Geographic, October 26, 2010

19. Not only was the traditional aristocracy out of touch with the basic ecology of the land. The disdain for the field amplified the ethnic and pseudo-religious sentiment promoted by the Doctrine. See: What the Founders Really Thought about Race: The Racial Consciousness of U.S. Statesmen, by Jared Taylor, The National Policy Institute, January 17, 2012.

20. The natives saw it differently. In fact, their story remains one of the epic indigenous traumas of human history. And, by the evidence of today's climate disaster, the threat to the future of human survival is real. See the reflections of Steve Newcomb in: "Five Hundred Years of Injustice": The Legacy of Fifteenth Century Religious Prejudice, by Steve Newcomb, 1992. See also: Doctrine of Discovery: A Scandal in Plain Sight, by Vinnie Rotondaro, The Trail of History, September 5, 2015.

21. Read it in their book: <u>The Spirit Level</u>; Why More Equal Societies Almost Always Do Better by Professors Richard Wilkinson and Kate Pickett, (Penguin, March 2009). They are from the Equality Trust, and produced an informative lecture video titled Inequality, the Enemy between us?

22. Ibid, Wilkinson and Pickett

23. The work, which one performs in the process of getting that produce to the finished stage, at which point the produce, (or service), is ready

for market is known in Economic language as: "value-added". Tis entity includes the wages paid for the necessary work accomplished in the process by the responsible employee; Incidental costs to support the employee's work and the profit claimed by the employer.

24. In his paper: "How income inequality hurts America", Steve Hargreaves points out that since the 2007 recession ended, "growth has averaged just 2.2%. That compares to the 3.3% historical average since the Great Depression". Meanwhile, the median wage earner in America took home 9% less in 2012 than in 1999. See: "How income inequality hurts America", Steve Hargreaves @hargreaves CNN, September 25, 2013.

25. The Economic Policy Institute says: "In 2006 five Nobel Laureates and six past presidents of the American Economic Association joined hundreds of other economists in called for raising the minimum wage, finding that a higher minimum wage "can significantly improve the lives of low-income workers and their families, without the adverse effects that critics have claimed." See: Economic Policy Institute, "Hundreds of Economists Say Raise the Minimum Wage," available at http://epi.3cdn.net/88c6aac4ee16915866_ldm6iie1l.pdf.

26. Figure 1, above, gives an idea of how the minimum wage has been over the past several decades. Note the minimum wage of 25 cents per hour in 1940 would actually buy $3.40 worth of merchandise today. Ibid, Dave Jamieson for the Huffington Post, (06/02/2012). The NRA further falsely assured Congress that: "most servers already earn well above the federal minimum wage of $7.25, and that raising the tipped minimum wage could hurt kitchen employees and others who don't work for tips".

27. Wendell Berry describes the new era in: A Continuous Harmony, by Wendell Berry, c1970, 1972, p. 12

28. The diamonds mined in South Africa, (along with many other metals), come from areas so deep in the Earth's mantle that human technology has not allowed us to penetrate to such levels. It is so hot at these levels that our best equipment would simply melt before it penetrated that far. With such movement routinely occurring at these levels, one cannot expect anything buried in the Earth to remain intact where it was buried.

29. It was called a "cap and trade" program, designed to overcome the weaknesses of Kyoto's voluntary approach by relying instead on market incentives and penalties. Accordingly, the cap-and-trade idea proposed that governments would set ceilings on maximum allowable CO2 emissions - the cap - for a given set of polluting industries. Then, for every ton of CO2 that a polluter reduces under the cap, it is awarded one "permit" to pollute. Permits could be bought, sold, traded or banked for the future. Any plant that cut its emissions below the mandated level could sell its excess allowances to over-polluters. Over-polluters could buy these indulgences and keep on polluting. But over time, governments would ratchet down the cap, restricting allowances. This would drive up the cost of permits. Dirty plants would face rising costs of buying permits to keep operating, discussed in: "Green Capitalism: The God That Failed", by Richard Smith, Truthout News Analysis, 9 January 2014.

30. Ibid. Richard Smith.

31. Under capitalism, growth and jobs are more often than not at odds with environmental protection. The profit motive gets in the way of Earth stewardship. Ibid. Richard Smith.

32. The International Energy Agency (IEA) has concluded that, "Coal will nearly overtake oil as the dominant energy source by 2017 . . . without a major shift away from coal, average global temperatures could

rise by 6 degrees Celsius by 2050, leading to devastating climate change", according to a report: "Climate Disruption Depression and 2013 Emissions Set New Records" by Dajr Jamail, Truthout, 17 November 2014. Against this backdrop, one letter to the states, which champion extensive voter suppression laws predicts that the new Congress after the 2016 Presidential Election would "put an end to the war against coal". These revelations are part of a report entitled: "Conservatives urge states to 'fiercely resist' EPA's climate rule," by Laura Barron-Lopez, December 5, 2014

33. The content of this chapter is based on a DVD report: <u>The Crumbling of America</u>, c2008 A&E Television Network.

34. Stephen Flynn, in his novel: <u>The Edge of Disaster</u>, presents a well-documented case for the country to urgently address the overall neglect of the infrastructure. See <u>The Edge of Disaster</u>, New York, Random House, c2007.

35. This 2009 report from the American Society of Civil Engineers was updated in 2013 and the overall grade had been upgraded to a D-. It was the first time since the report had been published in 1998 that the grade had improved. According to the Engineers' estimates, the cost of bringing the maintenance level up to "an acceptable level, a total investment of $3.6 trillion is needed by 2020. That estimate was up from approximately $2.2 trillion quoted in 2009. The good news is that the Federal government had increased its allocation for maintenance from $72 billion to $2.0 trillion since 2009. See ASEC's 2013 Report Card for America's Infrastructure, by Doug Scott, c2014, online at: www.infrastructurereportcard.org.

36. Not only is a strong combination of infrastructural components the foundation of effective urban development, but keeping them in place and well maintained allows an economy to grow and prosper.

See: <u>The Roads that Built America</u>, The Incredible Story of the U.S. Interstate System by Dan McNichol, c2003, www.danmcnichol.com/

37. The new NY Bridge Project is being carried out by the New York State Thruway Authority. The project description includes a photo of the model, map and schematics of the scheduling of the project. See: <u>The New NY Bridge Project</u>, copyright, New York State Thruway Authority.

38. The Sacramento levees collectively pose a major problem in Central California and the Sacramento Delta. Tempered by a long history of significant levee failures, many farms and tracts of land have had to be abandoned, due to levee failures. A study documents several cases between 1930 and the present: <u>Sea level rise and Delta subsidence</u> - the demise of subsided Delta islands, by Jay R. Lund, the Ray B, Krone Professor of Environmental Engineering, University of California, Davis, 9 March 2011.

39. On 1-2 May 2010, The Nashville area experienced a record rainfall of 13.57 inches. The previous 2-day record of 6.68 inches set on September 13 and 14, 1979." The record for the entire month of May, 11.04 inches, was set in 1983. The 2010 storms broke that record in just two days. Section researched and written by Will Thomas, Archival Assistant, Tennessee State Library and Archives, 403 Seventh avenue North, Nashville, TN 37243, email: <u>exhibits.tsla@tn.gov</u>

40. The Engineering News Record carried a story on the project, which was about 74% completed at the time of the report: Engineering News Record, ENR.com - 29 February 2012.

41. Mr. Haskins also complained that at the time of construction, the engineers did not build the foundation deep enough to provide the

strength necessary to make the Dam safe. See report: Foundation Flaws Make Kentucky's Wolf Creek Dam a High-Risk Priority, by Luke Abaffy, Engineering News Record, 02/27/2012.

42. Commenting on the challenge of preparing for the inevitable, Stout said: "My worst nightmare is that the dam and a seismic event like New Madrid would happen at the same time." Emergency Response Teams ready for Wolf Creek Dam Disaster, a FEMA Report, 17, August 2007.

43. Charles Oldham is a local geologist, who identifies local, national and international geological experts, who, like him advise commonsense. He suggests in the "Readers Comments" at the end of the Engineering News Report above: "The CORE has not yet learned that you cannot engineer away the forces of nature." Ibid. Engineering News Record, Charles Oldham, PG, 25 July 2012.

44. Dam removal can take on a whole different course to the river. This is an account of an actual project: "World's Largest Dam Removal Unleashes U.S. River after Century of Electric Production" by Michelle Nijhuis, for National Geographic, August 26, 2014.

45. It was not supposed to happen. However, a massive coal sludge spill exposes the Tennessee Valley Authority to a major toxic sludge spill, making it "a poster child for the failures of self-regulation", according to the report: "Toxic Sludge Leaks Expose True Costs of Coal", by Kelly Hearn, 4 February 2009. This article appeared in the February 23, 2009 edition of The Nation.

46. The water and sewer mains can be huge, carrying enormous amounts of fluid. A sudden break on a long-neglected main can be disastrous! The Maryland Christmas break surprised its Montgomery County residents. In an interview with Nancy Floreen,

County Supervisor, she explained: "the 66-inch pipe was under a major highway and could have been deadly, had it happened during the rush-hour." The incident was contained in a DVD report: The Crumbling of America, c2008 A&E Television Network.

47. Jim Neustadt, Field Engineer for the County Commission also reported on the above breaks, to the DVD report: Ibid, The Crumbling of America.

48. Sanitary sewer back-ups into homes and other living spaces presents a major health hazard. Furthermore, low income families are disproportionately affected by such hazards, often tempting the victims to unduly risk family health, because they cannot afford the immediate expense of a more permanent solution. Where the storm and sanitary sewers are combined back-ups can be even more acute, especially where heavy rainfall is possible. Sometimes, children have to be sent home when heavy rain causes a back-up onto a school yard! Communities and school districts have varying levels of capacity to respond appropriately to such surprises. Visit the US Environmental Protection Agency web site: Combined Sewer Overflows (CSO) Home: For information on the CSO national program, please contact Mohammed Billah at billah.mohammed@epa.gov

49. Unknown water leaks can have other disastrous consequences, as reported by Denis Rubin, Fire Chief, Washington D. C., in his description of his crew facing two raging fires, only to find the adjacent fire hydrant dry on both occasions. Ibid, The Crumbling of America.

50. The leaks release 15 to 35 million gallons of water a day, depending on the amount of water the aqueduct is carrying. The solution was to build a bypass tunnel around the significant leaks, and grouting closed the smaller leaks in Wawarsing. Site work began in January

2013 and construction is expected to continue through the year 2021. A detailed report can be seen in: "Mayor Bloomberg, Deputy Mayor Holloway and Environmental Protection Commissioner Strickland Visit Upstate Construction Site as Critical Repairs to the Delaware Aqueduct Begin." Official Report released by the Mayor's Office of the City of New York, 4 November 2013. Contact: Marc La Vorgna / Jake Goldman (212) 788-2958 Chris Gilbride / Ted Timbers (DEP) (718) 595-6600

51. Ibid, The Crumbling of America. Visit the US Environmental Protection Agency web site: Combined Sewer Overflows (CSO) Home: For information on the CSO national program, please contact Mohammed Billah at billah.mohammed@epa.gov

52. Ibid, The Crumbling of America. Visit the US Environmental Protection Agency web site: Combined Sewer Overflows (CSO) Home: For information on the CSO national program, please contact Mohammed Billah at billah.mohammed@epa.gov

53. This is the best estimate of the status of the country's infrastructure as of 2013, courtesy the American Society of Civil Engineers.

54. When a group of concerned citizens got together to feed hungry victims of the 1906 San Francisco Earthquake, little did they guess that they had started an urban garden movement that would grow to serve dozens of community gardens throughout the San Francisco Bay Area, a century later. Now known as the San Francisco League of Urban Gardeners, (SLUG). Read further in: "San Francisco Legitimizes Urban Farming", by Catherine Adams, Fog City Journal, January 27, 2012.

V

❖ ❖ ❖

Back to the Garden!

Can we Restore Earth to a Human-Friendly Habitat?

TODAY, EVEN WITH our modern lifestyle, we are still supported and predominantly powered by Earth's systems and cycles. We are what Earth allows us to be. Our yearning for freedom in the spirit of personal independence without responsibility courts global-scale disaster. This phenomenon has been a constant recurrence throughout much of the past century in the post-colonial global community.

The most critical challenge facing Earth today is the global climate change, which is destined to leave our environment uninhabitable for future generations. Earth lives in a harmonious flow of systems and cycles, which continually gravitates toward the balancing of its interwoven cycles. The continuous balance is what makes Earth's lifestyle sustainable.

Our mission of critical environmental degradation issues a clear and present warning, which is already forcing communities to evacuate their homes as oceans rise. These climate refugees add to a global refugee crisis of epic proportion. A product of the colonial era, the refugee crisis constitutes the results of the European conquest of native lands, displacement of the indigenous peoples and consequent degradation of the natives' respective economies. The conquest also severed the relationship, which native traditional societies had developed with Earth through the stewardship of their land.

The degradation of colonial ecosystems includes such environmental contamination as what was voiced by Benjamin Franklin, way back in 1786, with his concern about lead getting into the water we drink[1]. Thus, the clash between the concept of industrial entitlement and the common good has been a continuous struggle.

A genuine interest in the "common good" is only possible, when personal freedom is coupled with responsibility and the claim to entitlement is prohibited. Traditional societies understood the necessity for mutual responsibility within freedom. Modern societies are yet to fully comprehend how basic this understanding of mutual support is to its long-term survival.

Earth has provided us with all of our daily needs. Earth has endowed us with a wealth of diverse resources, which can only last with the practice of sharing for the common good. We are individuals and we are community. We are individuals: not independent, but interdependent *in* community. We are free to fulfill our desires, within the context of sharing Earth with others in equal freedom. In this way, the sustainability of available resources is possible and becomes mutually accessible to everyone once more.

Modern Social Standards are Recent in Human Civilization:

Jared Diamond, in his book: <u>The World Until Yesterday</u>, develops a rich discussion on the attributes of traditional society that would be worth incorporating into our present lifestyle. He points out, with vivid relevance, using appropriate examples, the value of reconsidering the fact that most of today's lifestyle is based on the last 5,000 years of human history. That is less than 1% of accountable human history.

Far from being a romantic nostalgic discourse, his examples demonstrate that traditional society was distinctly more violent. Yet, because social structures were much more intimate, the intrinsic value of that human-scale existence, was that forgiveness and reconciliation flowed naturally to break cycles of violence. Thus, the society was able to control that violence, precisely because the violence was generally among people, who would have to continue to live together. Thus, reconciliation was always close at hand and heavily relied upon[2].

In today's society, the enormous size of urban settlements had incorporated a level of anonymity, which has increased anonymous violence in recent years. Although society in general has become less violent, some societies are approaching the level of civic violence comparable to that of traditional society without the propensity to heal the wounds inflicted by such violence. Diamond attributes the general reduction in violence to the modern custom governments instituting more objective law enforcement, which discourages the ancient practice of individuals taking law into their own hands.

Diamond's research has encompassed the globe. However, most of his work has focused on New Guinea, where stone-age communities are being brought together with the modern world. Such an inter-relationship has allowed for opportunities to see how modern and ancient practices can co-exist amicably[3].

Earth has provided us with a human-friendly habitat. Our very life - as well as our livelihood - depends on our willingness to share Earth's resources with all of life on the planet. Traditional society was anchored in these principles. According to their living standards, every living organism on Earth was understood to contribute to Earth's livelihood and the livelihood of other organisms. Thus, they saw themselves as being *all connected to each other in life.*

Modern society has been programmed to see oneself as an isolated individual in a sea of other isolated individuals. In reality, especially in urban society, we are utterly interdependent, depending on thousands of other individuals, who work to produce the myriad goods and services for daily use. Furthermore, modern society leads us to be oblivious to this crowd of servers in our lives, we are also oblivious to the fact that most of these workers are often underpaid, overworked and living in chronic poverty.

Our Chronic Togetherness:

We breathe! That is the first necessity in life. The air we breathe is provided for all life. The air we breathe is also shared. It contains a mixture of elemental gases. Our bodies take from the air what it needs and return to the air what it does not use. All this is done automatically and involuntarily!

In similar conditions, we drink recycled water from a common atmosphere. Our bodies are 80% water, which is continually being extracted from our skin by the atmosphere to make up the water vapor, which makes up the clouds that send back rain to the land.

We also need each other. Humans are social animals. We have spiritual needs, which attract us to each other. Parents raise children in family units, based on an instinctive need to reproduce, protect infants in ways that eventually prepare them to become self-sufficient. Such self-sufficiency depends on developing basic social skills, which allow and empower each of us to support each other in mutual respect and dignity within the family and the community.

Some Elements Critical to Appreciating Traditional Society: It is helpful to see certain critical traditional society limits, when compared with modern society. Traditional society existed mostly in a rural, village setting. They were small intimate groups, territorial by disposition and constantly alert to potential threat of neighboring groups. Thus, the allied neighbor today may become an enemy next week. As such, the concept of: "love your enemy", took on immediate practical significance in long-term community building and measuring temporary enmity.

We learned to enjoy and build relationships, which sustain society. We accepted and encouraged governance as an ideal framework, within which we could nurture civil society. Each of us, as our ancestors appreciated, brings something positive to society, which contributes to a level of peaceful co-existence that maintains an ongoing sense of the mutual togetherness.

The prevailing spirit of "live and let live" grows out of the basic recognition that we need each other. Civil society is the product of such a spirit – a sense of being, in the hope for widespread wellbeing. Most species of animals live and let-live. Mothers are natural protectors of their birth offspring. Among animals with a strong social structure, certain adults often take on the duty of caring for groups of infants and adolescents. Wolves are particularly adept at this type of behavior. Ground squirrels, pelicans, gorillas, cats and penguins also practice such societal nurturing. In fact, parental nurturing is the norm throughout the animal kingdom.

Early traditional society valued a sense of communal nurturing. Neighborhood and neighborliness are natural parts of early human society. People simply looked out for each other, even in combative societies. Mutual cooperation merely makes the effort easier and more fruitful.

One of the drawbacks of modern society has been that the economy has usurped the health of planet as the perceived top priority of global leadership. The current leadership cannot even agree on the fact that an uninhabitable Earth will not support any economy!

Another regrettable development has been the devaluation of a natural sense of mutual interdependence with other members of the animal kingdom. Even more regrettable has been the modern attitude of disrespect for Mother Earth itself. Yet, most regrettable has been the assumption that only domination systems can govern large countries.

The attitude of entitlement is particularly important to the dominant elite. Thus, the rest of society is offered up as a human sacrifice to their convenience.

CHAPTER 22

—❖ ❖ ❖—

The Tangled-Web Syndrome

And the Disintegrating Community

THE IDEA OF competition as the first instincts guiding inter-personal relationships is revolutionary in human history. It tends to disintegrate, rather than build community. It encourages a closely guarded system of segregated segments of society. Laws are enforced unequally.

Such is the structure of European "democracy", it is contrary to the natural codes of communal survival. In such a setting, the competition-based community lives in constant contradiction to its own goals of neighborliness and peaceful co-existence. Since we are actually connected, as we attempt to exercise "independence", we create the syndrome of the "tangled-web".

To complicate matters, the domination system actively practices systems of deception in order to control the population at large. The consequences to such deception affect the long term capacity of that population to self-govern. In extreme cases, deceit becomes clearly evident, curtailing the basic trust necessary for effective government.

One may remember being taught in kindergarten: "What a tangled web we weave, when we practice to deceive"[4]. The statement was meant to teach children to be truthful and honest.

The early European migration to the Americas was an extension of their feudalism. Indentured servants were hired to make the trip across the Atlantic with the prospective landowners who hired them.

They travelled together, with the landowners treating the indentured servants like their personal slaves. Most of the servants had never sailed before, which added to the natural animosity and suspicion of European society[5].

The indentured servants did earn the privilege of a social status above the natives and later the African slaves. That privileged would later fuel the rise in Jim Crow laws and validate the White Supremacy movement. It continues to justify contemporary practices of displacing the natives and occupying reservation land.

The Declaration of Independence of the USA: It was written very much in the context of eighteenth European society. The founding fathers were in the hierarchy of a British colony. Their primary purpose was to become independent from Britain. Thus, the Declaration was written in the context of colonial hierarchy.

The United States of America began as a self-governing former colony. Today we look at the Declaration of Independence in the context of today's modern social customs. Its universal appeal is embodied in the words: "All men are created equal, with certain inalienable rights: liberty, equality and the pursuit of happiness". Notice: there is no mention of responsibility for the common good, the corner-stone of genuine democracy.

Moreover, at the time of writing, "men" were understood to be exclusively colonial gentlemen. Excluded were women, children, slaves, indentured servants and anyone else, who was not in the hierarchy of the "landed gentry".

The Declaration belonged to a colonial domination system, coated with the sense of entitlement of "manifest destiny". Its nature of persistent domination has survived the lip service of denouncing the Doctrine, while proclaiming the Millennium Development Goals of the United Nations, adopted by many religious institutions. The many carefully worded documents avoid associating the Doctrine of Discovery with the abject poverty and landlessness of indigenous peoples in the Americas and throughout the world[6].

The Tangled Web in Perspective:

Earth's lifestyle seems to exude harmony in the interplay between the four cycles, which we have been studying. Traditional societies seemed to take naturally to supporting mutual efforts to establish and maintain patterns of self-government with an emphasis on consensus-building. Like us, they appreciated that self-government had value. Unlike us, they valued it as the cement holding community together.

A strong economy needs community. Economic growth depends on strong markets. The US economy, once the strongest on Earth has thrived on consumerism. It was strongest in the 1960s and 1970s, when its middleclass was strongest. Income was relatively well distributed and people believed in the idea of democracy. However, the idea of democracy had been built on the spirit of feudalism.

Colonialism is a feudal system, built on the pyramidal structure of kingdoms. In it, the idea of democracy was always a stretch. The US Declaration of Independence was based on the glory of feudalism recited among the men of the feudal aristocracy of the early industrial revolution. The US independence and the French Revolution had much to celebrate. Yet, neither had achieved what both were celebrating. In tow was a grinding drag of wretched poverty being experienced by the majority of their respective populations.

Home on the Plains: In the meanwhile, the real wealth of North America lay on the plains. Jefferson saw it, but only partially, because European leaders at the time still saw the world from the top of a pyramid, the Roman way. They saw real estate and what it would fetch on a European market. The people of the Plains lived with herds of bison and managed them to feed millions without the squalor of the millions in the cities of the East Coast. They had a thriving middleclass.

They lived in tents and could carry what they owned on foot or horseback. But, they had what they needed. They lived on the banks of large rivers on the open plains, in tornado country. Yet, they escaped the damage that tornadoes could create when towns and cities were built to provide a target. The herds of bison were the wealth of the plains. They fed well on the lush, extensive grasslands and they supported the native population in relative prosperity.

Today, this is Tornado Alley. Its rivers drain the "bread basket of America", transporting the overloaded nitrate runoffs and other chemical fertilizers from the farms to kill the once teaming aquatic life of the Gulf of Mexico, with the aid of pollution from the industrial accidents of the petroleum industry.

The modern lifestyle has taken us away from such awareness. However, the animals have kept their vigil of stewardship. One can still see this pattern being played in the wild today. It is as if a particular animal is the designated warning agent for a specific danger. Like a neighborhood trait, the alarmists in one particular area may be different from those in another. A local grass fire ignited by lightning could arouse the prairie dogs, which may alert a flock of birds. Others would become acutely aware of their surroundings, testing their own sense of danger and act according to their own customs, (see related discussions in Section II).

In times of other extreme catastrophe, predator and prey animals may huddle together to protect themselves from the greater danger of a flood. In a major flash-flood that devastated one of the main valleys fed by the Mt. Pinatubo volcanic eruption, for instance, surviving villagers reported themselves peacefully huddling on rooftops next to snakes and rodents to escape the rushing mud of a lahar, (see Chapter 8 on volcanic operations).

The Urban Disconnect: Urban descendants of traditional natives in many catastrophic natural events may suffer disastrous consequences, because the relevant customs of their ancestors had been denounced by colonial leaders. Thus, in the wake of the 2004 tsunami in Southeast Asia, the natives who had moved to the urban areas were helpless, while their primitive cousins in places like rural Andaman Islands were unharmed. Casualties among urban natives were in the thousands, (see Chapter 15).

In addition, colonial "education" included attitudes toward animals regarded as "dumb" and "helpless", having nothing to teach us. After generations of ignoring local animals, people seldom recognized the warnings of approaching major natural events[7].

Hierarchy is Irrelevant to Earth's Lifestyle:

First of all, the importance given to hierarchy in the domination system is silly! What makes men automatically superior to women? The living victims of major catastrophes should be consulted by those seeking to solve the causes of such disasters. They seldom are.

Actions speak louder than words, and the way we approach the biggest challenge of our times speak loudly to our true intentions. Genuine leadership in our community comes from those, who seek to engage and involve as many useful hands and ideas as are available.

Furthermore, if we want government by the people, for the people, we must admit that all the people *are* people. Thus, they should all be represented at the policy-forming and decision-making tables. Poor people are well-equipped to decipher and work out sensible solutions to problems related to chronic poverty. They should have input into proposed solutions to their problems and challenges. Otherwise, genuine solutions are not being sought.

Our current leaders are usually the least equipped to seriously address many of the problems of governance. When we look at the dilemma faced by managers responsible for the infrastructure, (Chapter 21 – Crumbling Infrastructure), the field inspectors are seldom the critical decision makers. This is why so many obvious options are overlooked at critical staged in the ongoing maintenance cycles.

The European Version of Democracy is not Democratic:

Based on the domination system, the contradiction is built-in. Majority rule is essentially programmed to ignore, or devalue the opposition. It is a combative system designed to win arguments and defeat oppositions.

Community problems are more often solved by "all hands on deck" approaches.

In contrast to "majority rule", the ruling elders of traditional society, at the point of decision-making, kept all ideas offered alive on the table, only to be eliminated by mutual consent. The emphasis was not on winning; but on the collective consent of the best alternative. This is one of the values of traditional society that should be re-applied to modern society.

"By the people, for the people" in a natural, Earth-centered setting is meant for all people, unconditionally. This traditional society instinct is essentially community-building. The spirit is focused on people *working together* for the common good. In knowing and trusting the idea that our similarities always outweigh our differences the participants in a consensus system of governance come to trust all their colleagues as integral equal partners in effective self-government.

The Benefit of Consensus over Majority Rule in Government: Majority-rule advocates tend to respond to objections to their opinion as a reason to "fight". A common response to the advocates in favor of addressing the climate crisis is to "fight global warming". The emphasis on winning directs a struggle toward generic victory.

In contrast, the role of consensus-building is to work together for a unanimous outcome of the specific issue at hand. When leaders seek group decisions through consensus, it is clear to all concerned that the primary goal is for all to buy into the process of working together for the common good.

The irony is that issue-oriented combat focus often loses sight of the issue at hand and its relevance of the "common good". The focus is often inclined in favor of securing victory at all cost.

The US distrust for long-range planning often opts to apply short-term quick-fixes to long-term challenges, with the will to win over-shadowing support for solutions that favor the common good. This shortcoming was a common theme in addressing the Crumbling Infrastructure dilemma, (Chapter 21).

The short coming of issue-oriented combat as a basis for government of the common good is that important issues may never focus, if a popular issue takes center stage. A case in point is the current election campaign for the President of the USA. At a time when the need for a positive response to the global climate crisis by the US is most urgently, three debates have failed to address the challenge with any substance[8].

The Colonial Legacy in the Context of Climate Disruption:

The current dilemma facing today's humanity is a climate disruption crisis, which threatens the future of life as we know it. As the warming trend continues, extreme drought threatens food production in some areas. Water shortages reduce agricultural activities. Warmer climates in critical areas produce insect pest infestations that consume critical vegetation. On coastlines, the sea is rising.

Mining activities in the mountains continue to contaminate soils, and water tables downstream, precisely because the decisions to mine seldom involve the residents in the mining region. Mining activities displace large quantities of rocks and other materials of little importance to the mining operation. Accordingly, unstable deposits of debris are generated and often become hazardous to nearby residents.

Less obvious at the mining site is the potential impact the mining activities can impose on the watershed affected by the mines. Large-scale mining inevitably involves the displacement of large amounts of ground water. As such, a dam is necessary to contain the water, since it would be contaminated by the minerals and related chemicals exposed in the mining process. As seen in the coal ash dams discussed in Chapter 18, these dams are generally built as temporary, substandard structure. A dam is necessarily built in a drainage basin. Its contents is poised to contribute to some natural watershed. In effect, the activities of the mine are destined to impact the entire natural drainage basin downstream from that mine.

The prevailing tradition of poor safety management and protection of public places naturally vulnerable to mining activities is well documented. Poor countries, which are generally inexperienced with management of industry, are at even greater risk where mining is concerned. Global international regulating bodies are not protecting these poor countries. In a planet with an acute and growing shortage of clean drinking water, the global water shortage is made even more acute.

As discussed in Chapter 19, privatizing water supply is a direct conflict of interest. Furthermore, since air and water are a living necessity, there is a moral obligation to prohibit the commodification of the resource. A country's water belongs to the people it should remain a public resource. In addition, the availability of clean water to the public is a health and hygiene imperative. Where drought becomes a product of the climate crisis, the availability of clean water is rendered that much more critical.

Industrializing Colonies: The above observations, show some aspects of the colonial legacy that have adversely impacted former colonies. What is seldom appreciated is the reality that the people of these respective lands have been forced into an industrial lifestyle without their opportunity to choose such a lifestyle. The natives had an agreeable way of life, working for themselves and each other.

They lived in harmony with nature, planting food to eat with neighbors, who would grow or craft other products, which they could exchange for the extra food they grew. As a community, they fed and clothed each other and provided what else they needed among themselves. But, most of all, they were stewards of Earth.

As stewards, they wasted nothing. For the Plains people, the Bison was a huge staple. After the meat, the hide made moccasins, saddles, a wide variety of containers and pouches. The fat from the animal was used to make soap and cooking oil. Bones could provide knives, arrowheads, shovels, winter sleds. From the stomach came cooking pots, buckets, cups dishes. The hair supplied fillings for pillows and cushions, as well as, rope, head dresses.

As the railroad went west, so did the Europeans. There were 60 million bison on the plains in 1800. The hide and fur trade paid $3 per hide. Some hunters could kill up to 250 a day! They would skin them and leave the carcass to rot in the field, while the natives watched in dismay. Bison lived in herds and sometimes they would wander onto the railroad tracks. The railroads would pay to have them shot and removed. This led to a new sport of shooting bison from the train as it rolled through.

Naturally this led to many skirmishes between the immigrants and the natives. The US Army solved that nuisance by killing off bison to punish the natives. In 1889, William Hornady, Superintendent of the National Zoological Park in Washington, D.C. wrote a report accounting for 200 bison left in the wild[9]. To appreciate the resource waste represented here, picture the management of a few million bison would have kept native managers occupied and still supplied millions of hides. And, the plains would have been reserved for thousands of harmless tornadoes[10].

Accordingly, throughout the world the colonial movement displaced millions of people, who had lived in close harmony with Earth's lifestyle. Their descendants make up much of the current global refugee population. The crisis is too visible to be ignored and the younger global leaders, who are emerging are taking a more sober look at the refugee dilemma. Through the stories told by the history of the dilemma long-term significance of incompetent colonial systems of government can be realized. We can borrow from these lessons and improve on the future living standards of many communities.

An Emerging Global Concern for Future Viability:

Fortunately, the European countries have taken more seriously their responsibility to serve the common good. Because their public education has been more effective than the US in bringing their people more realistically into the 21st century, they were ready and able to respond more appropriately to the challenges of climate disruption, when it surfaced as an acute global crisis.

The Danish government was first and most immediately decisive in its official commitment to become independent of foreign oil. Following that lead, The Netherlands embarked on their own drive to increase their reliance on alternative energy. Germany, in the wake of the Fukushima disaster made the bold move to phase out their nuclear power plants.

Some communities, such as ReGen Village in a suburb of Amsterdam, in the Netherlands, are already building the viable alternatives for the future from existing technology. This community has committed itself to adapt to Earth's lifestyle by applying the best and most appropriate modern technology to an Earth stewardship lifestyle.

The ReGen Village experiment has already taught us how to use household compost to feed livestock and supply an entire aquaculture system. Completely self-contained, the village recycles its own waste to support a fish pond, while using fish-waste to help fertilize the growth of fruit and vegetable. Water is recycled to redistribute clean drinking water, as well as various stages of gray water.

Fresh fruit and vegetable will be grown in an aquaponic system, using one grade of gray water, much of it in roof gardens. Energy will be generated from geothermal solar, wind and biogas energy. Other gray water will be recycled for irrigation and secondary household use. The aquaponics system will deliver 10 times the yield of conventional farming, with 10% of conventional water use[11].

Apply Earth's Lifestyle to the Way We Treat Each Other: The most urgent challenge for future generations will be to develop a genuine democracy based on consensus. We have the collective wisdom to accomplish the transformation to Earth's Lifestyle, neighborhood by neighborhood. Much of the discussion contained in the genuine forces behind the Arab Spring and the "Occupy Movement" in the US, called out for more, as well as, more effective representation. We have to create forums in many settings to engage more people in focused discussions designed for public input from a personal voice.

We collectively know how to do this. One such process was aptly demonstrated at the Paris Accords on Climate Change in December 2015. The popular media has been quick to say that it cannot work and that is partially true. We cannot achieve consensus by continuing to use the combative majority-rule approach. Our very approach to governance *has to change!* In effect, we must commit ourselves to community building by consensus. We have to honor all people – the entire human race, without exception. The United Nations has taken on part of this challenge with their "Achieving Universal Energy" project[12].

As we saw in Chapter 21 on the crumbling infrastructure, when we are not honest about the reliability of what we have built, we often facilitate disaster. We are so comforted by the immediate present that we become indifferent to actual future outcomes. On a larger scale we have built the disastrous environment within which we currently struggle to survive. Accordingly, if we seek to reverse the trend toward rising global temperatures, we cannot expect to do so while preserving the current lifestyle.

A genuine transformation from our customs to a new way of approaching life is needed. It is not easy, but vital. Without such a transformation, we perish! A window of time exists within which we should act.

Widespread Earth Stewardship Can Transform Economies:

The process of reducing carbon emissions has spawned a growth in the use of alternative energy production. The resulting transformation requires the development and application of new technologies to deliver the new products.

In the political arena the transformation has already begun: The 2010 United Nations Environment Programme, (UNEP) produced a

report of eight projects which had successfully supported environmental integrity. Entitled "The Green Economy", the UNEP defined its
standard as: "one that results in improved human well-being and social
equity, while significantly reducing environmental risks and ecological
scarcities". These programs are summarized below:

China: When China began to implement its 11th Five-year Plan
(2006-2010), they allocated a significant share of investments to green
sectors, with an emphasis on renewable energy and energy efficiency.
Passed in 2005, China's Renewable Energy Law committed itself to producing 16 percent of its primary energy from renewable sources by
2020. Incentives to comply with the Law were immediately effective.
A national fund fostered renewable energy development by providing
discounted lending and tax preferences, which encouraged major advances in the development of both wind power and solar power.

By 2011 China had become the world's largest market for solar hot
water, with nearly two-thirds of global capacity. More than 10 per cent
of Chinese households rely on the sun to heat their water from total col-
lector area of more than 160 million square meters. Regular availability
of hot water has also brought with it considerable health and sanitation
benefits.

Being the largest solar panel manufacturer in the world, China produced 45 percent of global solar panels in 2009, rapidly becoming a
major market in Asia and the rest of the world[13].

Kenya's energy consumption was predominantly met by using traditional biomass energy, especially in rural households. Cities and urban
settlements maintained a heavy dependence on imported petroleum.

In March 2008, faced with high and unstable oil import prices,
Kenya's Ministry of Energy adopted a Feed-in Tariff, hoping that
"Renewable Energy Sources (RES) including solar, wind, small-hydro,
biogas and municipal waste energy would provide income and employment generation, over and above contributing to the supply and di-
versification of electricity generation sources". The Feed-in Tariff (FIT)
is a policy instrument that makes it mandatory for energy companies

or utilities responsible for operating the national grid to purchase electricity from renewable energy sources at a pre-determined price that is sufficiently attractive to stimulate new investment in the renewable sector.

Accordingly, the FIT would ensure that "those who produce electricity from identified renewable energy sources such as solar, wind and other renewable sources have a guaranteed market and an attractive return on investment for the electricity they produce. Aspects of a FIT include access to the grid, long-term power purchase agreements and a set price per kilowatt hour (kWh)"[14]. Initially covering wind, biomass and small-hydro, the policy was planned to include geothermal sources of energy. The boundary between Kenya and Uganda includes the Rift Valley, which is punctuated by several geothermal sites.

Uganda: In 1994 a few enterprises began a deliberate focus in organic farming. By 2005, 85% of the country's workforce was engaged in agriculture production, contributing to 42 percent of the national GDP and 80 percent of the exports earnings. Farming methods also focused on making their farming sustainable to improve efficiency. By 2003, Uganda had the world's 13th-largest land area under organic agriculture production and the most in Africa. By 2004, Uganda had around 185,000 hectares of land under organic farming covering more than 2 percent of agricultural land, with 45,000 certified farmers.

By 2005/06, 85 percent of the population was engaged in the agricultural sector. As the global market for organic foods and drinks grew steadily, the people had prepared to participate effectively. The global market was estimated to be around US$50 billion, and increased by 10-20 percent annually between 2000 and 2007.

Brazil has the fourth-largest urban population after China, India, and the US, with an annual urban growth rate of 1.8 percent between 2005 and 2010. The city of Curitiba, capital of Parana State in northern Brazil has implemented successful, innovative urban development systems in recent years. Its bus rapid transit system has inspired other cities in Brazil, and beyond. Equally impressive has been Curitiba's example of integrated urban and industrial planning that enabled the location of new industries and the creation of jobs.

These innovative successes have been timely for Brazil, which has been challenged by uncontrolled growth in residential and business expansion. The country can now hold up Curitiba as a model for other municipalities to follow[15].

India's National Rural Employment Guarantee Act, (NREGA), of 2005 is a guaranteed wage employment program that enhances the livelihood security of marginalized households in rural areas. It has been designed to directly touch the lives of the poor and promotes inclusive growth, while contributing to the restoration and maintenance of ecological infrastructure.

From 2006 to 2008, NREGA generated more than 3.5 billion days of work reaching on average 30 million families per year. All 615 rural districts in the country have been affected. Women represented roughly half the employed workforce, with emphasis on labor-intensive work, prohibiting the use of contractors and machinery. Along with supplementing wages, the program also hopes to strengthen rural natural resource management. This is achieved by financing rural works that address causes of drought, deforestation and soil erosion, thus restoring the natural capital base on which rural livelihoods depend[16].

Nepal: About one-fourth of Nepal's national forest is now managed by community forestry user groups, called CFUGs, which make up more than 35 percent of the total population. There are about 14,000 CFUGs, making them the second-largest forest management regime after government-managed forests. They develop their own operational plans, set harvesting rules, set rates and prices for products, and determine how surplus income is distributed or spent.

There is growing evidence of significant improvement in the conservation of forests (both increased area and improved density) and enhanced soil and water management. These groups seem to be improving the forest ecology, since being given this responsibility, which include employment and income generation from forest protection, tree felling and log extraction, as well as non-timber forest products. Additional economic benefits come from the sustained wood fuel sources, which contribute more than three-quarters of energy needs to households. Improved forest management and forest cover also contribute to nature conservation[17].

Ecuador: The city of Quito offers a leading example of the potential for developing markets that channel economic demand for water to upstream areas from which it is supplied. This form of watershed management focuses its attention on two ecological reserves in the Andes, the Cayambe-Coca (400,000 ha) and the Antisana (120,000 ha), which supply 80% of the city's water.

The municipal government, together with a non-governmental organization, have formed a trust fund to which water users in Quito contribute. The Fund for the Protection of Water – FONAG – was established in 2000 to finance critical ecosystem services, including land acquisition for key hydrological functions.

Water users pay differentiated rates depending on use. The largest share of payments comes from the Quito water utility (Metropolitan Enterprise of Water and Sewer Systems in Quito – EMMAP-Q) which contributes 1 percent of monthly water sales. Other users include farmers, hydropower companies, industries and households.

FONAG finances both watershed management projects in micro-river valleys and longer-term programs (at least 20 years in duration) oriented towards communication, environmental education, forestry, and river basin management training. Different community actors participate including local authorities, educational institutions, and governmental and non-governmental organizations. More than 1800 people are estimated to receive increased economic benefits associated with watershed management and conservation.

FONAG has served to inspire the development of similar schemes elsewhere in Latin America and beyond. For example, in South Africa, where water forms one of the greatest constraints on development, a recently-launched initiative in the Maloti Drakensberg Mountains aims to implement a payments for watershed services program, with support from UNEP and the BASF Social Foundation[18].

Tunisia: The government is preparing to promote the development and use of renewable energy, which would reduce the country's dependence on oil and gas. In 2005, an "energy conservation system" was established to provide energy management. This was financed by the National Fund for Energy Management, created to support increased capacity in renewable energy technologies and also improved energy efficiency. The fund is replenished by a duty on private, petrol-powered and diesel powered cars and by other selected duties. Between 2005 and 2008, the initial investments of $200 million in clean energy infrastructure had already allowed the government to save $1.1 billion in energy bills.

Successes among Low-income Communities: These successes have been localized. The benefits have usually been experienced by those affected, almost immediately. Probably the most visible of these impacts have come to the explosion of farmers markets in the past few decades. In fact, some of today's farmers markets have been operating continuously for many decades.

More recently, new markets have emerged in the inner-city of many metropolitan areas. These are unique in that they have been built on the examples of farms cultivated on inner city vacant building lots. These often are areas in which the local population has been deprived of reasonable access to fresh fruit and vegetable. These are poor people, whose children are often mal-nourished. For them, fresh vegetable are a ready source of inexpensive protein and critical vitamins[19].

Earth stewardship is an exercise in democracy, which can eventually guide many to the sustainable task of living in harmony with nature. It takes a few committed groups to stand up and claim the democracy of Earth stewardship.

The Netherlands present an example of what an engaged democracy can do to push for action on actively reducing carbon emissions, promised by its government. The European countries, and, indeed, most of the educated global population has a conscious awareness of world geography. These countries had signed on to the 2007 United Nations Intergovernmental Panel on Climate Change, (IPCC) protocol, which called on countries to cut back on carbon emissions[20]

Stichting Urgenda, a Dutch Foundation, was created in 2007 to promote a transition toward a sustainable and enduring society. In 2012, the Foundation grew impatient with the Dutch government for failing to act to meet the European Commission's climate goals. These include a target of 25 to 40 percent reduction in greenhouse gas emissions by 2020 in order to maintain a 50 percent chance of limiting global warming to 2 degrees Celsius[21].

Sustainability in Perspective: Like the Dutch, the Danish are responding to the needs of their citizens. Their population has been educated to understand and appreciate the reality of Earth's lifestyle. In fact, European awareness today acknowledges that global warming is happening and its effects are being written throughout the planet. The Danish government was awakened to the futility of their dependence on foreign petroleum for energy during energy crisis in Europe the 1970s. They created an energy commission, which decided to significantly reduce their dependence on foreign oil[22].

Each of us has a role to play in building the village we want for ourselves and our neighbor. An important task within the effort is to build a sustainable village, in defiance of the forces designed to keep us divided. Throughout the European region, governments have invested in mass-transit, with roads designed to also accommodate the safe use of bicycles for local urban transportation. In addition, many European countries are working with countries in Africa and Asia to help them install alternative fuels[23].

The Significance of the Climate Change Paris Accords: Although the US "experts" speculate that the Accord is weak, it remains clear that the world is concerned about the impact that the industrial countries are having on the planet. Its primary weakness is that it would have great difficulty moving forward within the existing framework of popular majority-rule government. Popular government is heavily reliant on enforcement mechanisms. The actual weakness is that it relies on force rather than developing community guidance through consensus.

Many small island countries have significant coastal areas, which are falling victim to the rising ocean. In this respect, an agreement signed on by 196 countries shows the universal consensus that we work together toward a significant reduction in carbon release into the atmosphere.

The interactions between the delegates at the Paris Accords suggests a growing sense of cooperative resolve among the nations. It is encouraging to see that, through an exercise in consensus-building the industrial countries consent to assist the developing countries[24].

Revive the Village for
Future Generations

It Takes a Village Attitude to Support Genuine Earth Stewardship

STEWARDSHIP IS A way of life, which acknowledges our genuine oneness with Earth and each other. It was clear to them that they all shared Earth in its totality. They shared the air they breathed and the water they drank, as well as the food that came from the land.

Traditional societies respected the power and providence of nature. They were simple people, for whom the idea of confronting or seeking to control nature was incomprehensible. They did not see insects as "pesky bugs", but appreciated that the insects fertilized flowers and seed to produce the food we eat, as did many birds and other wild animals. For them, "wild" meant *both* free *and* potentially dangerous. Thus, they observed each animal's way of life, respected it and gave it room to live its life.

They did not go to school, as we know "school" to be. Their environment was their school. Many were illiterate, by our standards; but by no means uneducated. By the time children reached puberty they often knew more about their environment than most of today's college graduates in the West. In completing the customary vision quest at puberty, they were fully prepared for functional adulthood.

Children's play was unplanned, spontaneous and related to their friends and their environment. Their skill development was generally relevant and complementary to the specific life skills of adults in their community. Eventually they would master some of those skills in adulthood.

These traditional villagers knew what crops to grow, when to plant and when to reap. They knew what to gather and when. They understood the soil and general habitat best suited for the growth of their favorite products. They saved everything and wasted nothing. For them recycling was as natural a way of life as sharing. As such, stewardship was the natural response to being alive.

The Utter Reliance on Earth's Lifestyle is Still True Today: Traditional village life flowed with a level of genuine freedom unknown to modern urban society. Because of the critical nature of the current climate crisis, it is urgent that we get into a stewardship/sustainable lifestyle. Today's climate patterns are unlike anything in past human history.

People think and talk about "fighting" global warming. We do not know this current climate. If there were something to fight. How could we know what to fight?! Our most urgent task is to get to know this climate thoroughly. That would take a global sharing of research, which is happening, in spite of significant opposition to the need to understand what Earth is doing.

Throughout the world, people had lived in traditional societies that had thrived for thousands of years in relative stability. They resided in small self-governing settlements, in which villages used various forms of consensus building as the preferred governance tool. Their focus was community cohesiveness for the common good.

In recent decades, regional economic entities have sought to strengthen their respective economies with varying degrees of success. In many ways this is seen in the eyes of the people in countries involved as an attempt to improve the lives and economic wellbeing of themselves and the member countries. On a separate page, corporations are competing behind the scenes to secure unfair advantages over others by incorporating clauses with special privileges written into the common agreement.

We Need an Earth-friendly Approach to Community Building:

We know that the major causes of the climate disruption are challenging the future living condition of the planet. We know the scale of the crisis. We know the disadvantages being faced by future generations, given the damage we have done to this point. We know what our relationship with Earth needs to be, in order to continue to be served on a sustainable human friendly level.

That we have not responded to the warnings and guidance we have had over the past two decades is evidence of our collective lack of responsibility for each other. From what one could expect in the next three or four decades, why aren't people concerned about their own descendants, or the state of the only planet on which they can live?

Manifest Destiny builds a community spirit of privileged entitlement. Leadership in the domination system openly celebrates the attitude and implies that every worthy person should be so blessed. In an affluent society, manifest destiny seems to promise a level of "universal entitlement" for those, who are successful in the society. In the U.S., the privilege of manifest destiny seems to be reserved for people of European descent.

However, for the popular concept of entitlement to work, a segment of society must be subservient and/or willing to work for a fraction of what one needs in compensation. Accordingly, an underlying attitude of entitlement is exercised by the privileged over the rest of society. Such a matrix of inequality is unsustainable in the long run, within a system that claims to be democratic. These built-in contradictions nurture an increasing sense of chronic injustice against the unprivileged under the law.

Accordingly, this system celebrates laws that are selectively enforced in this segregated society, a further contradiction to democracy. In such a setting, authentic democracy is not possible. If indeed, all men were created equal, the systems we live by are designed to contradict its own code and continue to undo the basic foundation of democracy.

The Writing on the Wall:

We had won the Cold War and are making the planet unlivable for human habitation in the process. That victory has taken us further along the road to human self-destruction. The present War on Terrorism is a huge mistake. It is time to cash in on the proposed "Peace Dividend" promised in the late 1980s, at the end of the Cold War. This was just one of the more recent warnings we did not heed.

The first point to be made here is that our majority-rule basis for leadership, being based on winning arguments, rather than consensus, is inclined to mislead and distract us from making wise decisions. For instance, in response to the alleged energy crisis of the late 1970s, President Jimmy Carter had 30 solar panels on the White House as a gesture to the nation that he was serious about the alleged energy crisis. When President Ronald Raegan was elected in 1980, one of the first things he did was to remove those solar panels[25]. Would a consensus discussion among the objective experts, removed from the pressures from Oil corporations, have led us to take seriously the global warming threat?

Riversong's timeline highlights dozens of such warnings in our collective awareness in my lifetime, since 1942. His timeline on such warnings begins in 1824, when Joseph Fourier explained the existence of the greenhouse effect!

It isn't just that consensus building is better for community-based governance. The more critical liability of exclusive reliance on majority-rule governance tends nullify the importance of most issues in the heat of battle relative to winning the argument at hand. President Raegan won that argument. But, had there been a public debate with consensus forums, how many groups would have turned down the possibility of reducing their local utility bills by hooking up solar panels to their homes in an energy crisis?

Other globally significant warmings include, (but are not limited to), the Hiroshima and Nagasaki bombings at the end of World War II, Followed by:

- Paul R. Ehrlich's book: the Population Bomb, in which he explained the greenhouse effect of increasing carbon dioxide, with low-level clouds concentrating contaminants in the air we breathe,
- In the 1960s smog was becoming a major problem affecting many cities globally,
- In 1972 the United Nations Conference on the Human Environment in Stockholm, Sweden raised the issue of climate change.
- In 1975 the world population reached four billion.
- In 1979, the World Climate Conference of the World Meteorological Organization concluded "it appears plausible that an increased amount of carbon dioxide in the atmosphere can contribute to a gradual warming of the lower atmosphere, especially at higher latitudes".
- In July of that year, The United States National Research Council published a report, in part corroborating the findings of the World Climate Conference. That was the year President Carter installed the solar panels on the White House.
- In 1998, James Hansen of NASA assessed that human-caused warming was already affecting the global climate in a testimony to Congress.
- That year the World Meteorological Organization established the Intergovernmental Panel on Climate Change, (IPCC) with the support of the UNEP, which issues a series of Assessment Reports.
- In 1989, Margaret Thatcher, based on her degree in Chemistry, warned about the vast increases in carbon dioxide in the atmosphere in a presentation to the UN.
- That year global carbon emissions from fossil fuels had reached 6 billion tons for the year.
- In 1990 IPCC produced its First Assessment Report, concluding a temperature rise of 0.3 to 0.6 degrees C over the previous century.

- In 1992 an international environmental treaty was negotiated in Brazil, which went into force in March 1994.
- In 1995 a Second Assessment Report discerned human influence on the global climate.
- The 1997 Kyoto Protocol agreed to a pledge by developed nations to reduce emissions by 5%by a period of 2008 to2012, with wide variations in target. The US refused to ratify the treaty.
- In 1999 the world population reached 6 billion.
- In 2001 President George W. Bush removed the US from the Kyoto process and proceeded to appoint a special committee to create research denying the validity of global warming.
- Following the bombing of the World Trade Center in New York in September 2001, President Bush declared a War on Terrorism[26].

Thus, the US public has tended to regard the climate crisis as just another arbitrary issue. This extreme has not occurred in Europe. They seem to value and respect academic expertise. In addition, the public seems sufficiently sophisticated to demand more of their media.

Public Education Matters:

Throughout the world, primary and secondary education include basic geography as a staple course. In the US, geography is usually an optional course, which is actually nonexistent in many school districts. As an instructor of physical geography at the community college level, I witnessed a steady decline in the knowledge base of recent high school graduates from 2002 to 2009, when I retired. The foreign students, even those with limited English, consistently scored higher than local students. The exceptions were US citizens, who had studied abroad.

The most telling trends include the gap in the domestic involvement of the public in steps to adapt living conditions to reflect more Earth-friendly customs. In Europe, Japan, China, India and many other parts of the world the expectations of the public reflect a greater willingness to sacrifice the convenience of the private use of the automobile and rely more on mass transit. More of their taxes are devoted to reductions in fossil fuel consumption. In general, the public response to the need to adapt to the demands of the current climate crisis was discussed in the previous chapter under the heading: "Widespread Earth Stewardship Can Transform Economies"

Probably the most striking indicator in Europe has been the government will to listen to their people and respond to their demands. The examples of Denmark and the Netherlands have been discussed above. We have elaborated on the disadvantages of majority-rule governance over consensus building. The gap is much greater in the US, where open hostility to environmental protection was supported and sometimes promoted by both President Raegan and President G. W. Bush while in office.

There is, nevertheless, a clear indication that the US public has privately supported a host of persistent and well organized opposition to the corporate powers been and often a popular media who often serves the corporate interests over the common good.

Back to the Writing on the Wall:

We have a tendency to think that "what's done is done." That is true on one level. On another level, I am influenced by an inspirational leader in my customer service training long ago. That is the level at which I want to present the partial list of opportunities above that may have significantly changed the course of history, had they been considered by a leadership of elders seeking consensus to guide them to act for the common good of the community they were assigned to serve. Let us consider a few of the warnings raised as alarms.

1. The Bombing of Hiroshima and Nagasaki was a sobering horror. It brought an end to World War II. However, it did not bring peace to the world. Why?

Europe went back to its gruesome cleanup, full of awful scenes of their bombed out neighborhoods. Similar civilian victims in Japan, the Philippines, Okinawa and myriad places throughout china, India and the rest of Southeast Asia and the Pacific Islands that had been host to the actual fighting of the war.

Indeed, the shell shock of that scale of war that the surviving soldiers, coming back from the war, would be carrying and sharing with old friends and other surviving buddies from the field. Those memories would not have allowed them the foresight to appreciate what it takes to revert to genuine peacetime living.

The returning combat survivors had no debriefing community support system to help them struggle through the post-traumatic stress symptoms, the awful scenarios of a buddy dying in his arms, or seeing his flight comrade's aircraft going down in flames, falling out of formation, who had been flying next to him only minutes before.

Peace was still perceived as a "break in the routine of war preparation as security". At many levels the atomic bomb awakened a new excitement in the dominant societies, who saw security in having one of their own. So we inherited the Cold War in place of peace and that peace opportunity was lost.

Then, there was the excitement of converting atomic and nuclear power to the peaceful supplying of domestic energy. However, the military was the lead in this effort. Military minds do not nurture peace. In effect, peace for its own sake did not surface as a predominant goal. Thus, it did not develop.

The Opportunities that were Missing: Millions of soldiers came back from World War II and never had the opportunity to debrief their traumas back at home with the drinking buddies. These were haunting memories someone would have great difficulty putting into words, simply because such feelings cannot be put into words. And men were taught from infancy not to cry. So, at this time, when their heart and soul wanted to weep, they couldn't!

And, the general public, who had been at home in the USA, so far removed from the actual combat, were delirious with the joy of a long war ended, could not take the time to be with the emotionally wounded soldiers, with traumas too deep for words. So, many of them were left to become helpless alcoholics, condemned for their illness, induced by the war we sent them to fight to keep us safe!

Missing was the time that loving community folk would take just to sit and be with them with patient, loving care to empathize and help them debrief. War had become greatly advanced to produce an atomic bomb. However, the psychology of post-traumatic stress disorder had not yet developed to help the returning soldiers, who had given their life for their country and were left in solitude to continue to live through the deep emotional cost of their sacrifice.

The postwar opportunity most missing from the period following World War II was reconciliation. That war should not have happened. But it did! And each of us harbored negative emotions, which, in the hearts and minds of others in our regional and global community, had led to that war. Hitler can be easily identified as the leader. But, he received considerable emotional support from the public. Local community groups, meeting to discuss how we contributed to the spirit of war that supported the action, would have helped the community to heal from the disease of war-mongering.

That type of healing is necessary to open up the opportunity to foster a sense of genuine peace in the community. In effect, we have to lay a tangible foundation for genuine peace. We have to cultivate fields of fertile harmony into our collective lifestyle, in order to live in harmony. This is the essence of Earth's lifestyle.

2. The 1972 UN Conference on the Human Environment: The next warning I want to address is the UN Conference on the Human Environment. This came on the heels of Paul Ehrlich's book: *The Population Bomb*. The Conference was undoubtedly was very much influenced by the public discourse that the book had ignited. Unlike the appeal for peace after World War II, this was more inclusively a public

debate. However, the debate did not get elevated into a consensus building process with the capacity to lead to inclusive public policies.

David Cameron, as prime Minister of the United Kingdom, wrote a proposal to the G-20 governments in the hope of building consensus among the leaders. It was a different type of consensus from what would be appropriate for building community at a grassroots level. Nevertheless, it appeals to the type of open dialogue necessary for building mutual agreement on group action[27].

As we become more adept at using consensus and phasing out majority-rule forums, we will find consensus more appealing, especially as a vehicle for establishing community cohesiveness. Such forums, applied in tandem with other community-enhancing activities will open up more spontaneous neighborliness. Furthermore, with the focus on the human environment, and the many relatively inexpensive sustainable options available, communities will break down the scourge of wide income disparity and come closer to eliminating destitution and reducing chronic poverty.

3. <u>We Never had an Energy Crisis</u>: The oil industry had an energy crisis. It was having difficulty cultivating their absolute control over the energy used in the economy. Earth has always had the sun, which provides far more energy than we would ever need. And our progressive scientific community has had the technology to harness more energy from the sun than we could ever exhaust. After all, stone-age man harnessed the sun to provide energy, starting with basic warmth derived from sitting in the sun on cold days, or choosing to cultivate fields in areas exposed to the sun to varying degrees to favor certain crops that thrived in direct sunlight.

Having solved the energy crisis, what is left is the will to phase out the fossil fuel alternatives, upon which we still rely. As technologies become more streamline and artificial barriers to the pursuit of solar and wind energy are eliminated. Utilities should facilitate customer transition to more independent harnessing of the energy customers require for home and small business needs. Customers still need the energy harnessed by thee utilities. In a free enterprise system, people should not be forced to purchase what they do not want, or need. If consumption purchase is forced, then the enterprise is not free!

The alleged energy crisis of the past four decades has done more to stifle economic growth, than to actually stimulate growth. The legal price-gouging of the petroleum industry, scandalized by the promotion of gas-guzzling SUVs and sheltering them from the real free-market competition of smaller, more efficiently operating automobiles, is a case in point[28].

The Energy Crisis and Climate Change Denial: In addition to withdrawing from the Kyoto process in 2001, President Bush facilitated and protected a consulting group: The Global Climate Coalition was committed to confuse the public into feeling that the global warming scare was a hoax[29]. As such, the public should disregard the alarm.

4. When President Bush declared war on terrorism, it was a warning of disaster in an arena too steeped in blind vengeance to appreciate the extent to which it was actually giving consent to uncontrolled self-destruction. One needs to unpack the declaration to appreciate what it means:

a. War *is* terrorism.
b. If one terrorizes terrorism, what are we left with?
c. Giving the task to Homeland Security implies that terrorism is lurking in our midst. How will it be defined? And by whom?
d. Arming local police with assault weapons further fuels a civil war without actually declaring it and without defining the battle lines.
e. Give the police the sense of entitlement to execute racial profiling and innocent civilians become the target of being hunted like criminals.

In a country, which honors and facilitates community consensus building the declaration of a war on terrorism would have been recognized as a dangerous, risky and futile endeavor and hundreds of innocent lives would have been saved. More critical, reasonable civility would have been restored soon after the bombing of the World Trade Center and we probably would have avoided the last ten years of war in the Middle-east at its current scale.

Re-establishing Human-scale Community Living: It cannot be over-emphasized that we are living in a war-torn human environment. The concept of a "war on terrorism" has invaded our consciousness. It has been driven home by the routine security checking in airports and many public buildings. The ready availability of firearms in the US means that firearms are also more easily available throughout the urban-industrial world.

As the "war on terrorism" has become a global pursuit, incidents of random public terrorism has multiplied. Furthermore, as populations have grown and settlements have become more crowded, each incident has often become more devastating. From a purely humane stand point, we need to make our communities more human-friendly. The public at large has to become more genuinely friendly. It is still the most acute societal need in today's society. The combative forms of government so popular and pervasive need to give way to more consensus building from the grassroots to local governing bodies. We must find ways to stop arming our settlements with deadly weapons.

For the sake of our grandchildren, do we want their future to be like this? Do we think this is an optimal living arena for them to raise their children? What is today's Presidential Election campaign offering them? Is that what we think of them? Are we satisfied with the image our country presents of each of us to the world?

CHAPTER 24

❖ ❖ ❖

Building Community for Future Generations

IN TODAY'S URBAN environment our modern lifestyle has conditioned us to mutually accept a low level of basic human dignity. Our crowded urban settlements have helped facilitate this conditioning. This has, in turn, led to a growing sense of insecurity, which seems to demand that we cultivate a heightened state of independence.

However, the same urban environment can only exist because it has built in an extremely complex infrastructure of interdependence. In effect, our lifestyle has created a matrix of customs that continually produce conflicting interactions. Thus, we need social customs that allow us to resolve conflicts regularly. If we hope to live in peace, then we need to cultivate forums that facilitate peaceful resolutions to conflict.

In traditional societies communities were sufficiently small to allow people to know everyone in their community. They had no independent police force. People tended to "take the law into their own hands", generally in hand-to-hand combat, which made their society often more violent than ours. That is not a good alternative and we do not want to return to such a standard.

However, because they knew each other, they developed, by necessity customs that facilitated reconciliation to heal conflicts. This is a custom we should consider incorporating into our lifestyle. That is one reason why we need to make it a priority to build community everywhere. The regular use of consensus strengthens the community-building process.

Consensus Building Breaks Down Alienation in Crowded Communities: Sixty years ago in the US, the local police were called "peace officers". They *were* community police. Today that phenomenon only exists by special arrangement. In Europe, the phenomenon lasted much longer and still exists naturally in many communities. Yet, each large city is made up of several smaller communities. This is a condition that needs cultivation in all the cities that would assist in revitalizing Earth stewardship.

Within this possibility, the opportunity to build effective community holds a key to making our settlements safer and more secure. A key to such success is developing local governance in neighborhood settlements around a consensus forum.

Communities are generally built from within. In effect, a community has to want to be inclusive, neighborly, the village intent on raising children with a healthy disposition. It also helps people to develop a caring disposition in their community, when it is in a country that encourages such development.

The US is not yet ready to provide leadership in caring for Earth in this crisis. Future generations will get the moral support they need from many influential groups in the US. The young leaders of tomorrow will receive more and better support from other parts of the world.

Many poorer countries are receiving technical support from European countries. Many are learning to adapt with low level technical assistance that they can maintain on their own over time. Those that can avoid military distraction are finding ingenious ways to adapt to their particular setbacks being brought on by the changing climate.

In the meanwhile, a major distraction to the task of removing the main causes of the climate disruption crisis. It is President Bush's War on Terrorism. The climate crisis was upstaged by the Bush war, which was kicked off by an attack on Iraq, based on his false claim that Iraq was hiding "weapons of mass destruction".

By the time the falsehood had been identified, the Government of Iraq had been dismantled and an alternative government imposed on the people of Iraq, which was never able to establish itself as the people's best choice. The misfit eventually gave rise to ISIS, the Islamic State of Iraq and Syria, as the upheaval in Iraq spread into neighboring countries.

Young Adults are Particularly Hard Hit by the "War on Terrorism":

Moreover, as our modern lifestyle becomes increasingly violent and the militarized police make themselves less accessible to the general public, urban life has become increasingly insecure in many parts of the US. Throughout this period of growing civil unrest, racial profiling by the police has intensified in many racially segregated area increasing the sense of civilian insecurity in the target population[30].

The War Mentality Engulfs Future Generations: This militarized violence is accentuated among young people, who have borne the brunt of police assault in recent years. The victims are not just African Americans in the U.S. We are sending more and more young people to war and abandoning then without proper psychological counselling. In this culture of war, violence has become a contagious disease!

A paper written by Henry Giroux, a professor in English and Cultural Studies at McMasters University in Canada mentioned 270 mass shootings in the US in 2014 alone. Other researchers, columnists and observers produce figures that would suggest that the US has achieved a war-zone scale of violence for a country that is supposedly at peace[31]. This is what we are preparing to pass on to the next generation.

This scenario suggests that too many youth in the US are living in a war-zone, while our most popular entertainment continues to celebrate the spontaneous use of hand guns and a culture of excessive violence. Yet, as if the violence did not exist, increasingly people are learning to become immune and ignore it. As modern society becomes more immune to the inhumanity of war, our already established war against nature is deepened, infringing on efforts to promote Earth stewardship

Nevertheless, young people of today have inherited an amazing array of promises and opportunities that seem endless. These opportunities are not available only to the privileged and others with a decent education. More and more people are finding in modern technology solutions to the impact of current climate disruption. Those, who have the will to offset current environmental degradation are recognizing that solutions exist. Their range of creative options are being used to make opportunities for themselves.

We Must Confront our Culture of Extreme Violence:

We have to abandon the aspects of our modern lifestyle that amplify punishment of the offenders, without seeking to heal the victims of crimes. Our customary court proceedings merely use the victim as a witness to serve the prosecution, seeking revenge against the offender on behalf of the state. Revenge only tends to escalate the vengeance and multiplies the pain and suffering of the victim and sympathizers.

The People of South Africa Chose Healing: The people of South Africa brought in a Truth and Reconciliation Commission, (TRC), after the end of Apartheid, to offer healing to a very volatile situation[32]. It has been a long process, but, the TRC and a community of compassionate people were sufficiently truthful to admit their collective pain and suffering. Thus, in 20 years they are healing.

Knowing this, why is the U.S. unwilling to heal? It is because the US "democracy" continues to seek resolve in the spirit of combat, trusting vengeance alone would satisfy the victim. Furthermore, they place ultimate reliance on the entitlement of the police judgment in their war against domestic terrorism to justify whatever steps they deem necessary to retain order and control. As we see approaching the US Presidential Elections of 2016, the rise in police brutality against African Americans keeps accelerating seemingly out of control[33].

In Canada, the Royal Commission on Aboriginal People's report, published in 1996, brought unprecedented attention to the residential school system. The report condemned the attempts of the residential program which sought to assimilate the native children into the Euro-American culture and eliminate the indigenous culture.

The University of British Columbia, through its Indigenous Foundation, assisted the Canadian Truth and Reconciliation Commission in seeking to heal the deep wounds inflicted by the Residential School System on the students. At first, an Aboriginal Healing Fund had received a $350 million government fund to aid communities affected by the residential schools in 1998. However, the native groups collectively protested the insufficiency of the fund and in 2007, the government agreed to fund the full amount necessary to effectively execute a new program to address the needs of all children affected[34].

In our system of acceptable extreme violence, the propensity to take care of Earth is subdued. Yet, Earth itself gives vivid examples of healing and spontaneous recovery. One has only to observe the tree in one's backyard recover and heal itself after a storm. We can learn a great deal about healing from watching a forest recover from a forest fire.

Healing the Community from the War on Terrorism: The alleged "terrorists" are often innocent victims of the war, itself. This is particularly true of profiled African Americans and Latinos. As such, it has been a source of immunized racism. Furthermore, the voices of certain police officers captured on their body cameras indicate an indignant sense of entitlement to abuse the profiled victims.

Necessary healing from this travesty requires healing of the entire community. However, many in the community would not entertain the personal compassion necessary to participate effectively in a corresponding reconciliation process. We have to want to work together to establish harmony in community, before healing would proceed.

In recent decades, survey researchers have devised special survey methods to create consensus decision-making in communities. One such organization is the International City/County Management Association, (ICMA). It organizes surveys and forums, designed to help groups to establish a measure of consensus in decision-making efforts. Other such agencies include Peak Democracy, a survey consultant in Berkeley, California[35].

The Promise of the Paris Accord:

A common assumption among the users of the majority rule format is that consensus decisions take too long. The Paris Accord experiment demonstrated the use of consensus decision-making among diverse interests in a large assemblies. It is unlikely that the majority-rule process would have been as fast at delivering an agreement.

Furthermore, the entire body of 195 nations celebrated the final decision in a way that is seldom seen in a majority rule process[36]. Consensus follows another natural phenomenon. Among people, there are far more similarities than differences.

Our Natural Tendency toward Cooperation: It should not be a surprise that cooperation is more natural than competition in building community. Most "natural disasters" demonstrate this phenomenon, as indicated earlier.

This human element is present in the lifestyle of traditional society. In our modern lifestyle, we tend to mistrust cooperation. We have taught ourselves to adhere to competition as a "natural" alternative. But, here again, we do not expect it to work, so we do not try it and it does not!

Reaching Back to Traditional Society: When the Paris meetings decided to use the indaba method of decision-making, it was at great risk. Indabas were first introduced in climate negotiation talks in Durban in 2011. In the last minutes of the meeting, negotiators reached a deadlock.

Indabas are used in complex negotiations among the Zulu and Xhosa people of Southern Africa, when seeking consensus on difficult issues, involving several parties. To prevent talks from collapsing this time, the program coordinators intentionally managed the process more closely. South African advisors to the process asked representatives from the main countries to form a standing circle and speak directly to each other. The entire process became a seamless adaptation into an old system of consensus building

Meetings were taking place in several rooms simultaneously, beginning on Wednesday evening. Some groups went on until the next morning. One went to 5 am, another until 8 am. Rooms had capacity for up to 80, with many standing around the room against the walls. Negotiators experienced in working with indabas were always available. Whenever an impasse emerged, key decision-makers would huddle in a corner of the room with negotiators for 30 to 45 minutes, to work out an acceptable agreement. According to veteran negotiators, the indabas made sure every country felt their views had been heard.

Parisagreement.org analyzed the task undertaken by the indabas. The Paris meeting began with 1,609 sets of disagreements. In the Wednesday night text there were 361. By Thursday night, it was down to 50.

Many Tools Available for Achieving Consensus:

As mentioned above, we already experience informal consensus in our daily lives. Our modern domination system relies on a level of informal consensus in order to work.

Surveys are another effective methods of finding consensus. However, the survey researchers have to be intentional in this task of seeking consensus. It is an active, participatory process.

We have participated in many opinion surveys, which pretend to be asking your opinion, when they are actually seeking *your agreement with their opinion*. Many market surveys and political opinion surveys are designed this way.

Our modern lifestyle is somewhat issue-oriented. We tend to use these issues to divide the community into warring parties. Unfortunately, in many circles, the environment and climate change has become just another issue to fight over.

The Most Vulnerable of Future Generations:

The most vulnerable of future generations are the children of today's most vulnerable people: the millions of refugees living in deplorable conditions throughout the world. These include the descendants of indigenous peoples globally, who are the direct victims of the Doctrine of Discovery. It was an amazing gesture, emerging from a post-colonial Christian movement

In February 2012, the World Council of Churches convened a conference in Bossey, Switzerland to study and reflect upon the complex impact of the Doctrine of Discovery on the global population of indigenous people. What they found was that the doctrine had rendered these people poverty stricken almost everywhere they had been forced to live throughout the world. In most cases, they were left "desperately trying to hold onto what they could salvage from their rich cultures and struggling to emerge from chronic poverty".

It would be appropriate to pause here to contemplate the plight of indigenous people. In most cases, these people had lived in their habitat for thousands of years before the arrival of Europeans. In North America, the refugees from Europe, who landed often sick and desperate, in vessels that were barely seaworthy, were fleeing religious persecution from England and Western Europe. The Europeans were in a land they did not know and the natives often had to assist them in making a living in their new habitat. By today's standards, driving their hosts off their own land was not the Christian thing to do, especially as they had nowhere else to go!

At the close of the conference of the World Council of Churches they had agreed to "repudiate the Doctrine of Discovery and to stand in solidarity with the indigenous people, in their continued struggle to receive the basic human rights, to which ordinary citizens are afforded,[37]. However, this was not the first repudiation of the Doctrine.

As early as 2007, the General Assembly of the United Nations had adopted a Declaration on Rights of Indigenous Peoples by a vote of 143 to 4, with 11 abstentions[38]. The USA, Canada, Australia and New

Zealand voted against the Declaration. This was a no-binding text, probably because in June 2006 a similar declaration was passed by the Human Rights Council with strong objection from countries with large indigenous populations[39].

Through the Eyes of the Indigenous: Consider the native victims of the Doctrine of Discovery. They were almost always people of a traditional society. They had lived in small communities in which everyone knew everyone else. They lived along the coast and on the banks of large navigable rivers. They had seen strangers arrive at some time in their own lives. They could have been traders from up and down the coast or the river.

Occasionally those along the coast had seen people arrive from Africa. They were generally friendly people, except for occasional attacks from unfriendly neighbors, with specific disputes. They had known their neighbors for centuries, through stories handed down by the elders and storytellers.

The Europeans were very different. They looked different and that aroused general curiosity. However, that, for the natives, did not pose an immediate problem. They had no idea that these particular visitors had arrived with a mission to enslave or displace them. It was a flood of immigrants like nothing they had ever seen and they occupied the land everywhere, with permanent houses. They spread over the land

Some 200 years later, the Europeans had reached the Mississippi River. They had moved unto the flood plains and taken over the land the natives used to plant during the growing season. They built villages and towns on the flood plains, so that we could no longer use them. The natives tried to tell them that the land flooded often, but they did not seem to care, until it actually did. Then, they would lose their crop and many of their prized positions in the houses.

The natives on the plains west of the Mississippi suffered some of the worst trauma, as they witnessed the slaughter of their herds of bison, which the refugees took only for their hides, or just for sport and left their future livelihood to rot out on the open field! The bison was the greatest wealth of the plains natives. It existed in the form of dozens of herds scattered over the plains, an estimated 60 million around 1800. It was painful to endure as the witnessed the loss of their bison staple[40].

Today, the North American natives are confined to marginal lands on reservations. Throughout the Americas, many parts of Africa, Australia, Asia, the Pacific Islands, the Caribbean and elsewhere, the descendants of native peoples languish in extreme poverty, the result of their ancestors having been driven off their native land. This has, in recent years, worsened the global refugee crisis. As such, many groups have become their advocates[41].

Existing Informal Consensus Groups: One of the most successful informal consensus displays has been the global denial of the impact of the Doctrine of Discovery on the global indigenous population. It is the proverbial large white elephant made invisible in the middle of a crowded plaza, there in plain sight. Thus, few seem to associate the Global refugee crisis with the Doctrine of Discovery and its lasting impact on the indigenous people living in former colonies.

Where similar community building efforts have been employed, they have had positive effects. Many community groups already operate very much by informal consensus. Community gardens and local neighborhood watch groups are successful living examples.

Other potential allies to efforts to develop greater use of consensus in local governance include a variety of cooperative organizations, such as housing cooperatives, farming cooperatives, some credit unions. In effect, many existing local groups are currently using consensus methods of community building, policy-making, and regular decision-making in their ongoing operations.

We Can, (Have to), Heal Our Own Wounds:

It takes a village to raise a child. It takes a village, committed to protecting the wellbeing of each other to raise a child, with the natural instinct to support and protect the wellbeing of others. It takes a genuinely democratic village to raise a child with the spirit and will to live in harmony with Earth's lifestyle.

Such a village lives by consensus and respects all other life, as a matter of course. The stewardship of Earth has been generally an integral part of village life in traditional society. The colonial lifestyle was injected into traditional society customs to contaminate Earth's lifestyle and replace it with a selfish, self-centered domination system, fueled by the industrial revolution to give us the modern lifestyle, which we crave and live for today.

That experiment sought to violate Earth's lifestyle and declare war on nature. Its mission has been to plunder Earth and exploit the "common good" for the exclusive convenience and enjoyment of a prescribed group of elite men in feudal society. It used an industrial revolution in order to create a global economy designed to maintain wealth within the chosen elite.

The colonial plunder of Earth has severely wounded the planet. Earth's wounds are everyone's wounds. Our current climate disorder is a symptom of Earth's woundedness.

The proposed stewardship of Earth is prescribed to enable us to reconnect with Earth's lifestyle as our own transformed lifestyle. The current climate disorder we are experiencing is Earth's way of healing itself. In becoming stewards of Earth, we can assist in that healing process.

Stewards of Earth soon come to realize that we are all mutually interdependent. As such, we all benefit from the positive relationships we cultivate with earth and each other.

We have to listen to Earth to appreciate its natural tendencies toward healing. For those who live close to the land, what stands out is Earth's resilience in the face of injury is its endless capacity to heal:

- Forests heal from the ravages of storms, as the sections of fallen trees die and regrowth takes place around the debris to fill in the gap created by the storm.

- Similarly, after a forest fires in a normal, unprotected forest, the scorched land, fertilized by the layer of ash, springs back to life with the first rain. Birds and other animals drop seeds and the next springtime is greeted with the green of new growth.
- After a major landslide, the scarred slope left behind eventually stabilizes the loose topsoil. New vegetation takes hold of the top soil and completes the healing, very much the way an injury to one's bruise heals.
- A flash flood erodes sections of hill slopes and the upper banks of rivers. As exposed soils stabilize, new growth hastens the healing of the remaining scars on the landscape.

We need to honor Earth's lifestyle and live in response to what Earth exemplifies. We are not used to living like that in modern urban society. Yet, "practice makes perfect" and one's commitment to stewardship becomes a meaningful incentive.

Some countries and communities are essentially committed to living by such a code. All such communities seek a measure of prosperity. People aspire to live in relative peace with their neighbor. Where this is not a temporary aspiration, sustainability becomes a high priority, which, when achieved carries its own reward.

In the Final Analysis: Earth stewardship as a daily practice would benefit everyone. For those committed to the practice, their effort would not feel like a burden, but more a sense of purpose. Active stewardship brings more meaning to life.

We belong to Earth. We should be grateful for what Earth provides for all of us to share. There is still enough to share, if and when we decide to share with more equity. Many of us can live with a lot less. It may take some getting used to; but it is eminently possible.

Thus, we must be vigilant in our pursuit of consensus democracy, in order to establish the harmony of Earth's lifestyle. Earth has promised that harmony for those who live in that spirit of "live and let live".

The ultimate question is: Are we willing to make the necessary sacrifice? The abnormal global climate changes we are currently experiencing stands as Earth's warning of the pending end of human survival. We have to pursue consensus democracy in order to appropriately engage our entire community in the greatest global challenge of our time – indeed of all times!

Notes

1. Concern for the environment is not new. As early as 1786 Benjamin Franklin warned about the public health challenge of lead and other

contaminants in public water supplies. See: Death Toll Rising in the Global War on Environmentalism, a collection of articles compiled by Word Press, ©2016.

2. People lived in small groups, generally of hundreds or thousands, with active codes of honor. Anonymity was virtually impossible. People lived their entire lives within a radius of a few dozen miles. They knew and were known to everyone in their community and often guarded their territory fiercely. There were frequent feuds. However, they also relied heavily on trade, because food was often subject to unpredictable scarcities. Water was not always easily available. Thus, they built-in several opportunities to reconcile differences with dignity, when situations of weather and uncontrolled inconvenience required groups at war to trade, or otherwise work together. See: Jared Diamond: The World Until Yesterday: What can we learn from Traditional Societies? Viking Press, New York, c2012.

3. The health and sanitation advances of the past 150 years are a clear advantage no one would offer to give up in favor of the corresponding standards of traditional society. However, these modern advantages do not require an abandonment of consensus democracy, or the value of family and community, ibid, Jared Diamond.

4. I was taught a ditty in kindergarten, which was to encourage me to speak the truth and not try to deceive others. I think it still works for me today. This chapter outlines a few of the issues confronted daily, in which public information has been designed to deceive. We observe the consequences to the public for following such misleading information. For example, consider the popular misconception that modern technology can improve on nature as played out among the innocent victims of most "natural disasters".

5. It is said that at the time of European conquest of the Americas: "the Holy Roman Empire was neither." The church was an intimidating force. What one claimed to believe could be grounds for torture. A glimpse at the Spanish Inquisition captures a sense of life in Europe at the time. See: "The Truth about the Spanish Inquisition", by Thomas Madden, Associate Professor, University of Saint Louis, ©October 2003

6. The United Nations Millennium Development Goals aim to "eradicate extreme poverty and hunger". It was supposed to have been halved between 1990 and 2015. Instead, extreme poverty has increased.

7. This phenomenon was also demonstrated in the 2004 tsunami, where an elephant was able to save a tourist family, in spite of the efforts of the native handler to stop it from bolting to safety. When they had a chance to consider what happened in the Andaman Islands in the face of the tsunami, the analyst was able to appreciate what the community, who had never left the forest had learned from their ancestors. They retreated into the forest and were all safe. In contrast, those who had become urban dwellers had lost contact with their ancestors' customs when they moved to town. Accordingly, many of them lost their lives in the tsunami. See: "Tsunami folklore 'saved islanders'" by Subir Bhaumik, BBC News, Andaman and Nicobar Islands, 20 January, 2005.

8. The debates have been so focused on one candidate, whose primary claim to competence for presidential leadership is the candidate's appeal to run the country as if it were a business. The climate crisis is openly denied and ignored. Thus, public discourse on its impact on the country and its people is not even addressed. Such a

development would not be possible in a consensus democracy. Too many people want to discuss the issue.

9. *According to Kathy Weiser's account of the period, one comes to appreciate how the vast wealth of North America was squandered and wasted to build the economies of the US and Europe. Weiser, Kathy, "Buffalo Hunters, Old West Legends", Legends of America, updated April 2015.

10. A 19th century North America living up to Earth's lifestyle would have left the plains as prairie land with some grain agriculture and other farming with low density settlements. Tornado Alley would have seen very few damaging tornadoes, because there would be little to damage.

11. The full account of the ReGen Village is outlined in the following report: This New Neighborhood Will Grow its Own Food, Power Itself and Handle its Own Waste by Adele Peters, ReGen Village, Amsterdam, May 25, 2016.

12. These "success stories" featured sites in Asia, Africa and South America. See: The Green Economy, Copyright © United Nations Environment Programme, 2010. For more information, see: www. unep.fr www.unep.org United Nations Environment Programme, P.O. Box 30552 Nairobi, 00100 Kenya. The purpose of their mission is better described in a later report: UNEP Report: Towards a Green Economy: Pathways to Sustainable Development and Poverty Eradication, 16 November 2011. These studies validate the general thesis of: Paul Hawken: Blessed Unrest, How the Largest Movement In the World Came Into Being and Why No One Saw it Coming, Viking Press New York, NY., c2007

13. Ibid, UNEP, The Green Economy.

14. Ibid, UNEP, The Green Economy.

15. Ibid, UNEP, The Green Economy.

16. Ibid, UNEP, The Green Economy.

17. Ibid, UNEP, The Green Economy.

18. Ibid, UNEP, The Green Economy.

19. Ibid, UNEP, The Green Economy.

20. Strichting Urgenda, a Dutch Foundation sues the government to honor its commitment to the IPCC: "Will judges force the Netherlands to meet E.U. climate goals?" by Erica Rex, E&E Correspondent, ClimateWire, March 12, 2013. The final version was passed on 21 May 2015 Contact Dennis van Berkel at: dennis.van.berkel@urgenda.nl

21. One has to appreciate the sense of urgency that the Dutch people feel, living behind formidable dikes built to hold back the North Sea. The ocean is rising and many Dutch citizens remember the Storm of 1st February 1953, when the North Sea engulfed the country. It is a story virtually unknown to younger US citizens. Yet, the Dutch lost over 1800, when the dikes were breached at high tide in that storm. It happened at night. They had no early warning system. Many awakened to a home flooded to the roof and simply drowned.

22. Denmark, as the first country in the world to do so, has decided to lead the transition and become a green and resource efficient economy entirely independent of fossil fuels by 2050. Reporting for the Danish Energy Agency the staff describes sustainable economic growth and security of energy and water supply, which are

among the greatest global challenges today. See: "IEA data point to emissions decoupling from economic growth for the first time in 40 years." By Dan Howis Lauritsen, 26 March 2015.

23. The story of small enterprises breaking through the barriers to solar energy education and exposure are finding a ready and willing audience in poor countries. Furthermore, the breakthrough is quite widespread, involving a variety of different technologies, which are easily applied. Read: "D,light Attacks a Big Energy Problem with Small Solar", by Marianne Lavelle, The Daily Climate, April 21, 2015. Several other articles have reported different international projects designed to help poorer countries to participate in efforts to reduce their dependence on fossil fuels: 42%: Danish Wind Power Sets World Record Again, by Anne Vestergaard Andersen, January 15, 2016. "South Africa is Ready to Unleash Significant Energy Efficiency Potential", by Nicoline Oturai, Bachelor and Helle Momsen, 28 April 2015. "Copenhagen no. 2 in Europe for Air Quality", by *Anne Vestergaard Andersen, 27 April 2015.* "Record High Exports of Danish Energy Technology", by Anne Vestergaard Andersen, 30 April 2015.

24. A unique process: the Durban Platform for Enhanced Action was used to create a consensus agreement out of the Paris protocol. Using a technique from the Zulu and Xhosa people of southern Africa they arrived at a landmark agreement among 196 countries at: The United Nations Framework Convention on Climate Change, (FCCC), Conference of the Parties 21st Session, Paris, 30 November to 11 December 2015. Durban Platform for Enhanced Action: Adoption of the Paris Agreement.

25. Robert Riversong has published a timeline on the world's awareness of global warming crisis: Riversong, Robert, "Turning the Tide: Shifting the Paradigm of Human Culture," December 2015, copyleft

by Robert Riversong: may be reproduced only with attribution for non-commercial purposes and a link to this page.

26. Ibid, Riversong

27. The initiative here is genuine, but would necessarily require consensus on many different levels, with forums designed to communicate effectively, back and forth between the different levels between the grassroots and the elected decision makers. See: Cameron, David, *Governance for Growth*: Building consensus for the future, Office of the Prime Minister of the United Kingdom, November 2011, ©Crown copyright.

28. In a study conducted by Friends of the Earth, a non-profit researcher. The tax loophole was described in the following paper: Moulton, Sean, "Gas Guzzler Loophole: SUSs and Light Trucks Drive off with Billions", with contributions from Brian Dunkiel, David Hirsch, Gawain Kripke, and Erich Pica, Design and layout by Annette Price, Graphic Designer. Friends of the Earth copyright ® 7000 www. Foe.org

29. The group: Global Climate Coalition was formed in 1989 in response to the United Nations Intergovernmental Panel on Climate Change, (IPCC). The group was disbanded, but was reinstated in 2002. Its earliest members included: Amoco, the American Forest and Paper Association, American Petroleum Institute, Chevron, Chrysler, Cyprus AMAX Minerals, Exxon, Ford, General Motors, Shell Oil, Texaco and the United States Chamber of Commerce. From the outset, their mission was to generate public doubt by discrediting the science of global warming. "Global Climate Coalition" resurfaced during the Bush Administration. Its research, written by the Center for Media and Democracy provided the corporate spin, which claimed that the climate change theory was a hoax, 12

January 2012. See also: Rosencranz, Armin, "U.S. Climate Change Policy under G. W. Bush, GGU Law Digital Commons, 32 Golden Gate U. L. Rev. 2002 http://digitalcommons.law.ggu.edu

30. When seen in the context of random gun violence the very prospect of personal security becomes increasingly dubious. The Centers for Disease Control and Prevention (CDC) states that "2,525 children and teens died by gunfire in [the United States] in 2014; one child or teen death every 3 hours and 28 minutes, nearly 7 a day, 48 a week." See: Gun Culture and the American Nightmare of Violence: "Focusing merely on mass shootings or the passing of gun legislation does not get to the root of the systemic forces that produced the US love affair with violence." by Henry Giroux, January 13, 2016.

31. Compared to US citizens killed in wars: More Americans Killed by Guns since 1968 than in All U.S. Wars, Columnist Nicholas Kristof Writes, by Louis Jacobson, August 27, 2015. A partial list of mass shootings prior to 2013 is reported in: A Timeline of Mass Shootings in the U.S. Since Columbine, by Aviva Shen, December 14, 2012.

32. We have seen the progress that has been made in almost 50 years, since the US embarked on its Law and Order movement of 1968. Review: More Americans Killed by Guns since 1968 than in All U.S. Wars, Columnist Nicholas Kristof Writes, by Louis Jacobson, August 27, 2015. Compare the South African experience: Truth and Reconciliation Commission, (TRC): South Africa's Commitment to Healing from the Apartheid Experience, 1995 to 2002. Produced for the South African History, Online. Last updated: 15 April 2016.

33. In recent years the domestic war on terrorism has focused on African American young men. The following studies speak to this issue: 1

Black Man is Killed Every 28 hours by Police or Vigilantes: America is perpetually at War with Its Own People, by Adam Hudson, AlterNet, May 28, 2013. One specific occurrence is reported in: A Response to Ferguson: Systemic Problems Require Systemic Solutions, by John Powell, Director, Haas Institute for a Fair and Inclusive Society, November 11, 2014.

34. When one group in a society is being arbitrarily targeted to be harassed by the police, the situation puts undue strain on a wider range of people in that society. The Canadian truth and reconciliation commission, from its experience challenging the country's past programs designed to erase the native culture from the lives of children, has much to share with the US. However, the US has to first openly denounce the war on terrorism as an unfortunate error of judgment and outlaw it in all its phases.

35. The International City/County Management Association, (ICMA), has conducted several such surveys worldwide, with encouraging results. See a sample approach in: *New Features Deliver Better Insights on Survey Feedback, by Marlena Medford, Product Updates, July 14, 2016.*

36. A unique process: the Durban Platform for Enhanced Action was used to create a consensus agreement out of the Paris protocol. Using a technique from the Zulu and Xhosa people of southern Africa they arrived at a landmark agreement among 196 countries: *The United Nations Framework Convention on Climate Change, (FCCC), Conference of the Parties 21st Session, Paris, 30 November to 11 December 2015. Durban Platform for Enhanced Action: Adoption of the Paris Agreement.*

37. To that point, the statement of the World Council of Churches had been unprecedented. Read the Council's official statement in:

Statement on the Doctrine of Discovery and its Enduring Impact on Indigenous Peoples, presented by the Executive Committee of the World Council of Churches, Bossey, Switzerland, 14-17 February 2012.

38. United Nations, "General Assembly Adopts Declaration on Human Rights of Indigenous Peoples; 'Major Step Forward' Towards Human Rights for all, Says President", (for information media – not an official document), 13 September 2007 - Sixty-first General Assembly, Plenary, 107th &108th Meetings (AM&PM)

39. Ibid, United Nations.

40. The herds of bison were the wealth of the plains people. There were about 60 million bison out on the plains around 1800. They were divided into dozens of herds scattered over the plain. Some herds had as many as one million animals. Indigenous people had a tremendous respect for life and did not believe in killing for sport. This means that when they killed animals, they took great care to use as much as possible, because it ensured the survival of the species they hunted. The hides were quickly skinned and tanned for clothing, shoes, blankets, teepee covers. Rawhide (the hair removed) was even more versatile as it could be used for making belts, snowshoes, moccasin soles, water troughs for horses or hide tanning, quivers, shields, buckets, drums and even rafts! The skin on the head of male buffalos was extremely hard. Native people often used it as a bowl. Tammy Robinson, "How Native People Used Every Part of an Animal for Survival," July 1978, © Copyright Off The Grid News.

41. Following the lead of the World Council of Churches, the Episcopal Church issued its own repudiation: Pastoral letter on the Doctrine of Discovery and Indigenous Peoples, by Katharine Jefferts Schori, Episcopal Presiding Bishop, The Episcopal Church Office of Public

Affairs, May 16, 2012. In July 2012, the United Church of Christ issued the following: "The Repudiation of the Doctrine of Discovery", by Elizabeth Leung, Ministry of Racial Justice, The United Church of Christ, 9 July 2012. Other bodies have expressed the overall need for reconciliation, including: "Doctrine of Discovery: A Scandal in Plain Sight", by Vinnie Rotondaro, The Trail of History, September 5, 2015.

Bibliography

A&E Television Network, *The Crumbling of America*, a DVD report, aired on A&E Television Network, c2008.

Alexander, Michelle, *The New Jim Crow*, The New Press, New York, c2012

Alexander VI, Pope (1431-1503), *Demarcation bull.*, granting Spain possession of lands "discovered" by Columbus.

Ambrogi, Thomas E., *1988 Progress Report*, the Institute for Food and Development Policy, Food First News, Volume 10, no. 35, Fall 1988.

Anderson, Carol, *Bourgeois Radicals*, The NAACP and the Struggle for Colonial Liberation, 1941-1960, published: December 2014.

Alperovitz, Gar, *America Beyond Capitalism*, Reclaiming Our Wealth, Our Liberty and Our Democracy, (paperback), 24 April 2006.

Anthony, Carl and Pavel, M. Paloma, *Breakthrough Communities*, Ecology Center, Berkeley, CA c2012

Augier, F. R. et al, *The Making of the West Indies*, Longmans, Green and Co., London, UK, c1960

Awatisgi, Nvnehi, (Jim PathFinder Ewing): *Finding Sanctuary in Nature*, Findhorn Press, Forres, Scotland,IV36 3TE, C2007

Baerwald, Thomas J., and Fraser, Celeste, *World Geography*, Building a Global Perspective, Prentice Hall, New Jersey, ©1998.

Baker, Dori Grinenko, ed,: *Greenhouses of Hope*, The Alban Institute, Herndon, Virginia, c2010

Ball, Edward, *Slaves in the Family*, Farrar, Straus and Giroux, New York, ©1998.

Barcott, Bruce, *The Last Flight of the Scarlet Macaw*, One Woman's Fight to Save the World's Most Beautiful Bird, Random House, ©2007.

Barlow, Maude, *The Free Trade Area of the Americas*, The Threat to Social Programs, Environmental Sustainability and Social Justice. Published by the International Forum on Globalization, February 2001, Volume I

Barrett, Samuel A. and Gifford, Edward W., *Indian Life of the Yosemite Region*, Miwok Material Culture, Bulletin of the Milwaukee Public Museum, volume 2, no.4 March 1933.

Behrman, Richard E., M.D.: *The Future of Children*, Long-Term Outcomes of Early Childhood Programs, Richard E. Behrman, M.D., Editor, Center for the Future of Children, ©1995.

Benor, Daniel and Harrison, James Q., *Agricultural Extension*: the Training and Visit System, for the World Bank, May 1977.

Berry, Thomas, *Dream of the Earth*, Sierra Club Books, San Francisco, ©1988, reprinted ©2015.

Berry, Thomas, *The Sacred Universe*, Columbia University Press, New York, NY., c2009

Berryman, Phillip, *What's Wrong in Central America*, and what to do about it, American Friends Service Committee, Orbis Books, ©1983.

Bolland, O. Nigel, *Struggle for Freedom*, Essays on Slavery, Colonialism and Culture in the Caribbean and Central America, The Angelus Press, Ltd., ©1997.

Boushey, Heather and Hersh, Adam S., *New Evidence in Economics*: The American Middle Class, Income Inequality, and the Strength of Our Economy, May 2012. www.americanprogress.org

Bowen, Ezra, *The High Sierra*, The American Wilderness, Time-Life Books, New York, ©1972.

Brinkley, Douglas, *The Great Deluge*, Harper Perennial, HarperCollins Publishers, New York, NY, c2006

Brower, David, *Let the Mountains Talk, Let the Rivers Run*, New Society Publishers, New York, c2000

Brower, Michael, Ph.D. and Leon, Warren, Ph.D., *The Consumer's Guide to Effective Environmental Choices*, for the Union of Concerned Scientists, Three Rivers Press, ©1999.

Brown, Michael P., (with Shari Prange): *Convert It*, a Step-by-Step Manual for Converting an Internal Combustion Vehicle to Electric Power, Future Books, Ft. Lauderdale, FL, ©1993.

Buntin, Simmons B., *Unsprawl: Remixing Spaces as Places*, a Case Study of Portland's River Place, Planetizen Press ©2013

Burt, Christopher C., *Extreme Weather*, A guide and Record Book, Norton and Company, ©2004.

Cabezas, Omar, *Fire from the Mountain*, The making of a Sandinista. Translated by Kathleen Weaver. Crown Publishers, inc. New York ©1985.

Cabot, Laurie, *Celebrate the Earth*, Delta Books, New York, c1994

California Coastal Commission, *California Coastal Access Guide*, University of California Press, Berkeley, CA., c1981

Camarda, Renato, *Forced to Move*, Solidarity Publications, San Francisco, ©1985.

Campbell, Bruce M. y Luckert, Martin K., Editados: *Evaluando la Cosecha Oculta de los Bosques*, Fundo Mundial para la Naturaleza, (WWF), ©2002.

Carson, Rachel, *Under the Sea-wind*, New York, Oxford University Press, c1941.

Carson, Rachel, *The Sea around Us*, Oxford University Press, c1951.

Carson, Rachel, *The Edge of the Sea*, Boston, Houghton Mifflin Company, c1955.

Carson, Rachel: *Silent Spring*, Boston, Houghton Mifflin Company, c1962, reprinted by RachelCarson.org c2002.

Carter, Brenda and Loeb, David, *A Dream Compels Us*: Voices of Salvadoran Women, South End Press, Boston MA ©1989.

Carter, Forrest, *The Education of Little Tree*, University of New Mexico Press, Albuquerque, MN, ©1976.

Carter, Jimmy, *Palestine Peace,* Not Apartheid, Simon & Schuster, New York, NY, c2006.

Carter, Stephen L., *Civility: Manners, Morals and the Etiquette of Democracy*, Harper Perennial, New York, NY, ©1998.

Chomsky, Noam, *Turning the Tide*: U.S. Intervention on Central America and the Struggle for Peace, Boston, South End Press. ©1985

Chomsky, Noam, *The Umbrella of U.S. Power*: The Universal Declaration of Human Rights and the Contradictions of U.S. Policy, New York, Seven Cities Press. ©1999,

Christopherson, Robert W., *Geosystems*, seventh edition, Pearson Prentice Hall, c2009.

Cooper, Matthew J.P., Beevers, Michael D., and Oppenheimer, Michael, *Future Sea Level Rise and the New Jersey Coast*, Assessing Potential Impacts and Opportunities Science, Technology and Environmental Policy Program, Woodrow Wilson School of Public and International Affairs, Princeton University, November 2005

County and Municipal Government Study Commission: *Housing and Suburbs, Fiscal and Social Impacts of Multifamily Development*, Ninth Report, October 1974.

Criddle, Joan D., *Bamboo and Butterflies,* From Refugee to Citizen, East-West Bridge Publishing House, Dixon, CA, ©1992

Dannin, Ellen, *Crumbling Infrastructure*, "Infrastructure Privatization Contracts and Their Effects on State and Local Governance," Northwestern Journal of Law and Social Policy, Cornell University Press, 2011.

Devall, Bill and Sessions, George: *Deep Ecology*, Peregrine Smith Book, c1985

Diamond, Jared, *Guns, Germs, and Steel*, The Fate of Human Society, University of California, Los Angeles, c1997

Diamond, Jared, *Collapse*, How Certain Societies Caused their Own Demise by Ruining the Environment that Sustained them, c2005

Diamond, Jared, *The World until Yesterday*: What can we learn from Traditional Societies? Viking Press, New York, c2012

Dillard, Annie, *Pilgrim at Tinker Creek*, Harper and Row, Publishers, New York, c1974

Dixon, Marlene, editor, *On Trial*: Reagan's War against Nicaragua, Synthesis Publications, San Francisco, ©1985.

Dobbs, Michael, *U.S. had Key Role in Iraq Build up*, Washington Post, Washington, D.C., ©2002

Dudley, Ellen and Seaborg, Eric, *American Discoveries*, Scouting the First Coast to Coast Recreational Trail, The Mountaineers, Seattle, Washington, ©1996

Duhon, David and Gebhard, Cindy, *One Circle*: How to Grow a Complete Diet in Less Than 1,000 Square Feet, Ecology Action, Willits, California, ©1985.

Earle, Ethan, *A Brief History of Occupy Wall Street*, Rosa Luxemburg Stiftung, New York, ©2012.

Edwards, Andres R.: *The Sustainability Revolution*, Portrait of a Paradigm Shift, New Society Publishers, c2005

Ehrenreich, Barbara, *Nickel and Dimed, On (Not) getting by in America*, ©2011

Evangelical Lutheran Church in America (ELCA), *Our Communities in Crisis*, ECLA Advocacy, May 2015, Washington, D.C.,

Everett, Melissa, *Bearing Witness Building Bridges*: Interviews with North Americans living and working in Nicaragua. New Society Publishers, Philadelphia, PA, ©1986.

Farmer, Paul, *Pathologies of Power*, Health, Human Rights and the New War on the Poor, University of California Press, Berkeley, California, c2003.

Ferlinghetti, Lawrence, *Seven Days in Nicaragua Libre*, City Light Books, San Francisco, ©1984, All Rights Reserved.

Figueres, José María, *From Decline to Recovery a Rescue Package for the Global Ocean* by José María Figueres, Trevor Manuel and David Miliband Global Ocean Commission Report 2014.

Flynn, Stephen, *The Edge of Disaster*, (New York: Random House, 2007)

Fortey, Richard: *Earth*, an Intimate History, Alfred Knopf, New York, NY., c2004

Fox, Thomas C., Iraq: *Military Victory, Moral Defeat*, Sheed & Ward, Kansas City, MO, c1991.

Francis, Pope, *Encyclical Letter: Care for Our Common Home*, Proclaimed by Pope Francis at the Vatican, 18 June 2015

Friedman, Thomas L., *The World is Flat*, Farrar, Straus and Giroux, New York, NY., c2005

Gargallo, Francesca, *El Pueblo Garifuna*, Cuadrenos Pedagogicos No. 18, Ministerio de Educacion, Guatemala, 24 de julio de 2002.

Gelbspan, Ross, Cullen, Heidi and Piltz, Rick, *Everything's Cool*, a video documentary on Official Attempts to Confuse the Public on the Science of Global Warming, Sundance Film Festival, ©2007

George, Susan, *A Fate Worse Than Debt*, Penguin Books, ©1988

Giroux, Henry A., *Hearts of Darkness*, Torturing Children in the War on Terror, Paradigm Publishers, Boulder, CO, c2010

Golden, Renny and McConnell, Michael, *Sanctuary:* The New Underground Railroad. Orbis Books, Maryknoll, New York. ©1986.

Government of Denmark, *The Danish Climate Policy Plan*, Toward a Low Carbon Society, August 2013.

Griffin, Michael and Block, Jennie Weiss: *In the Company of the Poor:* Conversations with Dr.Paul Farmer and Fr. Gustavo Gutierrez, Orbis Books, Maryknoll, New York, © 2013.

Grimes, Orville F., Jr., *Housing for Low-Income Urban Families, Economics and Policy in the Developing World,* a World Bank Research Project, ©1976.

Griswold, Eliza, *The Tenth Parallel*, Dispatches from the Fault Line between Christianity and Islam, Eliza Griswold, ©2010.

Gritziotis, George, *Mining Health, Safety and Prevention Review*, Ministry of Labour, Ontario, April 2015.

Hacker, Andrew, *Two Nations, Black and White, Separate, Hostile, Unequal*, a New York Times Bestseller, Scribner's Sons, ©1992.

Hansen, Eric, *Stranger in the Forest*: On foot across Borneo. Penguin Books, New York, ©1988.

Harewood, Jack, *Employment in Trinidad and Tobago, 1960*, Institute of Social and Economic Research, University of the West Indies, Jamaica.

Harrison, Bennett, *Building Bridges*: Job Training for Community Development in a Changing Job Market. A Report to the Ford Foundation, January 1995.

Hawken, Paul: *Blessed Unrest*, How the Largest Movement In the World Came Into Being and Why No One Saw it Coming, Viking Press New York, NY., c2007

Hawken, Paul, *The Ecology of Commerce*, a Declaration of Sustainability, Harper Publishers, Inc., New York, NY., c1993

Her Majesty's Stationary Office, *The Future of Development Plans*, a report of the Planning Advisory Group, ©1965.

Her Majesty's Stationary Office, *The New Towns of Britain*, Central Office of Information Reference Pamphlet #44, ©1969.

Her Majesty's Stationary Office, Trinidad and Tobago: *The Making of a Nation*, Central Office of Information Reference Pamphlet #53, ©1962.

Hill, Lawrence, *The Book of Negroes*, Winner of the 2008 Commonwealth Prize for Best Book. A Black Swan Book, Transworld Publishers, London, W5 5SA, United Kingdom. A Random House Group Company© Lawrence Hill 2007.

Ilibagiza, Immaculee: *Left to Tell*, Hay House, inc., Carlsbad, California, ©2007

Jeavons, John, *How to Grow More Vegetables*, Ecology Action of the Mid-Peninsula, 10 Speed Press, Berkeley, California, ©1982.

Johnson, Michael, coordinator: *Housing Technology for Model Cities*, a report on Housing Technology, for Metropolitan Detroit Citizens Development Authority, 1969.

Johnson, Michael, coordinator: *Operation Breakthrough Housing System Producers*, for the U. S. Department of Housing and Urban Development, February 1971.

Jones, Edward P., *The Known World*, Pulitzer Prize winner, ©2003, Amistad/ HarperCollins

Jones, Jacqueline, *A Dreadful Deceit*, The Myth of Race from the Colonial Era to Obama's America, Basic Books, New York, NY, ©2013.

Kaufman, Milton and Wilson, J. A., *Basic Electricity*: Theory and Practice, McGraw-Hill, ©1973.

Kidder, Tracy, *Mountains Beyond Mountains*, Random House Deluxe, New York, NY., c2009

Kristof, Nicholas D. and WuDunn, Sheryl, *Half the Sky*: Turning Oppression into Opportunity for Women Worldwide, Alfred A. Knopf, New York, NY., c2009

Kroeber, Theodora, *Ishi in Two Worlds*, University of California Press, Berkeley, California, ©1961, 1976.

Leiva, Julio: *Un Mejor Manana, El Terco Deseo de Crear*, Julio Leiva, San Salvador, El Salvador, C.A. c2005.

Lieberman, Gerald A. and Hoody, Linda L., *Closing the Achievement Gap*, Using the Environment as an Integrating for Learning Context for Learning, State Education and Environment Roundtable, ©1998, 2002.

Logan, William Bryant, *Dirt:* The Elastic Skin of the Earth, Norton & Co., New York, ©2007

Logan, William Bryant, *Dirt: The Movie*, The Sundance Film Festival, ©2009.

Lopez, Ian Haney, *Dog Whistle Politics*: How Coded Racial Appeals Have Reinvented Racism & Wrecked the Middle Class, ©February 28, 2014.

Louv, Richard, *The Nature Principle, Human Restoration and the End of Nature Deficit Disorder*, Algonquin Books, Chapel Hill, NC, c2011

Lowney, Chris, *Pope Francis*, Why he Leads How he Leads, Loyola Press, Chicago, ©2013.

MacEoin, Gary, editor: *Sanctuary*: a Resource Guide for Understanding and Participating in the Central American Refugees' Struggle. Harper and Row, San Francisco ©1985.

MacPherson, John, *Caribbean Lands*, a Geography of the West Indies, 2nd Edition, Longmans, Green and Co. Ltd., London, UK., c1967

Mails, Thomas E., *The Mystic Warriors of the Plains*, ISBN 0-7924-5663-7

Mann, Charles C., *1491*, First Vintage Books, New York, NY., c2006

Manz, Beatriz, *Paradise in Ashes*, University of California Press, Berkeley, California, c2004

Margolin, Malcolm, *The Ohlone Way*, Indian life in the San FranciscoMonterey Bay Area, Heyday Books, Berkeley, California, c1978

Mashable.com: *Wonderworld:* A photographic journey of planet Earth: "She may look beautiful, but Earth is in danger". By Mashable.com/2015/04/03/wonderworld-photos-of-earth/ April 3, 2015

Massachusetts Institute of Technology: *Solar Power Applications in the Developing World*, by Amy Rose, Andrew Campanella, Reja Amatya, Robert Stoner. An MIT Future of Solar Energy Study Working Paper, ©2015.

McGaa, Ed (Eagle Man), *Crazy Horse and Chief Red Cloud*, Four Directions Publishing, Pine Hill Press, Sioux Falls, SD, c2005

McIntyre, Alister, *Decolonization and Trade Policy in the West Indies*, The Caribbean in Transition, Institute of Caribbean Studies, University of Puerto Rico, Rio Piedras, Puerto Rico, 1965.

McKibben, Bill, *Deep Economy*, Times Books, Henry Holt and Company, LLC , New York, NY., c2007

McKibben, Bill, *Hope, Human and Wild*, True stories of living lightly on the earth, Hungry Mind Press, Saint Paul, Minnesota, ©1995.

McKibben, Bill, *Eaarth*: "Making a life on a tough new planet", St. Martin's Press, New York, NY., c2010, 2011

Merrill, Michael D., *Impounded Sediment and Dam Removal in Massachusetts*, A Decision-making Framework Regarding Dam Removal and Sediment Management Options, by Michael D. Merrill,

River Restore Technical Coordinator, Riverways Programs, MA Department of Fisheries, Wildlife & Environmental Law Enforcement, Boston, MA 02114, June 2003

Meyers, Judy L., et al: *Where Rivers are Born*: The Scientific Imperative for Defending Small Streams and Wetlands. The publication was funded by grants from the Sierra Club Foundation, The Turner Foundation and American Rivers, February 2007.

Ministry of Housing and Local Government: *Moving Out of a Slum*, a study of people moving from St. Mary's Oldham, London, Her Majesty's Stationary Office, ©1970.

Morris, Jim, *Slow-motion Tragedy for American Workers*, Lung-damaging silica, other toxic substances kill and sicken tens of thousands each year as regulation falters. A study sponsored by The Center for Public Integrity, updated: June 30, 2015.

Myers, Bryant L., *Walking with the Poor*, Principles and Practices of Transformational Development, Orbis Books, Maryknoll, New York, c1999.

Nazario, Sonia, *Enrique's Journey*, Random House Trade Paperback, New York, NY, c2006.

Nazario, Sonia, *La Travesia de Enrique*, Random House Trade Paperback, New York, NY, c2006,

Nerburn, Kent, *Neither Wolf Nor Dog*, On Forgotten Roads with an Indian Elder, New World Library, Novato, California, ©i994, 2001.

The New English Bible, Oxford Study Edition, Oxford University Press, c1961

Newcomb, Steve, *Five Hundred Years of Injustice*: The Legacy of Fifteenth Century Religious Prejudice, ©1992

Nicholas V, Pope, *papal bull Dum Diverseas, 18 June 1452*, issued to King Alfonso V of Portugal the bull Romanus Pontifex, which became the Doctrine of Discovery.

Norris, Kathleen, *Dakota, a spiritual Geography*, Houghton Mifflin Co., New York, NY., c1993

Obama, Barack, *The Audacity of Hope,* Thoughts on Reclaiming the American Dream, ©2006.

O'Brien, Mary Barmeyer, *Outlasting the Trail*: The Story of a Woman's Journey West, a Two dot Book, Helena Montana, ©2005.

O'Kane, Trish, *Guatemala:* a guide to the people, politics and culture, in Focus. Interlink Books ©2000.

Olson-Raymer, Dr. Gayle, *History 110, The Doctrine of Discovery*, Humboldt State University

Olkowski, Helga, *Self-Guided Tour to the Integral Urban House*, Farallones Institute, Berkeley, CA, ©1973.

Organization of American States: *Declaration of the Presidents of America*, Meeting of American Chiefs of State, Punta del Este, Uruguay, April 12-14, 1967.

Ortiz, Beverly R., *It Will Live Forever*: Traditional Yosemite Indian Acorn Preparation, Hayday Books, Berkeley, CA, ©1991.

Owenden, E. F. C. W. A.: *Costs and Earnings Investigations of Primary Fishing Enterprises*, a study of concepts and definitions, for the Food and Agriculture Organization of the United Nations, Rome, 1961.

Oxaal, Ivar, Black Intellectuals Come to Power, *The Rise of Creole Nationalism in Trinidad and Tobago*, Schenkman Publishing Company,Inc., Cambridge, Massachusetts, c1968.

Pacific Gas and Electric Company: *Resource,* an Encyclopedia of Energy Utility Terms, 2nd Edition, San Francisco, California, ©1992.

Pacific Gas and Electric Company: *Alternative Energy*, San Francisco, California, ©1984.

Pagels, Elaine, *Beyond Belief*, Random House, New York, c2003.

Pavel, M. Paloma, *Breakthrough Communities, Sustainability and Justice in the Next American* Metropolis, MIT Press, Cambridge, MA, c2009

Paul, Daniel N., *Doctrine of Discovery*: the doctrines decreed by Roman Catholic Popes: Source: First Nation History, Third Edition, 2006

Peck, H. Austin, *Economic Planning in Jamaica*, a Critique, Journal of Social and Economic Studies, Volume 7, 1958.

Preece, D. M. and Wood, H. R. D., *The Foundations of Geography, Book I*, 16th Edition, University Tutorial Press, Ltd., London, UK., c1967

Preece, D. M. and Wood, H. R. D., *Europe, Book III*, 8th Edition, University Tutorial Press, Ltd., London, UK., c1960

The President's Committee on Urban Housing: *A Decent Home*, U. S. Government Printing Office, Washington D. C., ©1973.

Prothero, Stephen, *American Jesus*: How the Son of God became a National Icon, Farrar, Straus and Giroux, New York, NY. ©2003.

Rechtschaffen, Clifford and Gauna, Eileen, *Environments Justice, Law, Policy and Regulation*, Carolina Academy Press, Durham, North Carolina, c2002.

Reynolds, Edward, *Stand the Storm*, a History of the Atlantic Slave Trade, Allison and Busby, Ltd., London, UK., c1985

Richter, Burton, *Beyond Smoke and Mirrors*, Cambridge University Press, Cambridge, UK., c2010

Ritchie, Mark Andrew, *Spirit of the Rainforest*, a Yanomamo shaman's story, Island Lake Press, Chicago, IL, c2000

Rodney, Walter, *How Europe Underdeveloped Africa*, Howard University Press, Washington, D. C., c1982

Romain, David, *First Five Year Development Programme of Trinidad and Tobago, 1958-1962*, a Critique, successfully presented as a thesis for M. A. in Geography, University of Durham, Durham, England, ©1969.

Romero, Archbishop Oscar, *Voice of the Voiceless*: The Four Pastoral Letters and other Statements. Translated from the Spanish by Michael J. Walsh. Orbis Books, Maryknoll, New York. ©1985.

Rotberg, Eugene H., *The World Bank, A Financial Appraisal*,

Rutter, Nat, Coppold, Murray and Rokosh, Dean, *Climate Change and Landscape in the Canadian Rocky Mountains*, The Burgess Shale Gepscience Foundation, Field British Columbia, ©2006.

Salzman, Jack, Editor-in-Chief, *The African-American Experience*, Information Now Encyclopedia, MacMillan Library Reference, ©1993, Charles Scribner's Sons.

Schlesinger, S., and Kinzer S,*Bitter Fruit*: The story of the American Coup in Guatemala, Cambridge, Mass., Harvard University, David Rockerfeller Center for Latin American Studies. ©1999

Sen, A., *Poverty and Famines*: An Essay on Entitlement and Deprivation, Oxford: Clarendon Press. ©1981.

Shimer, Porter, *Healing Secrets of the Native Americans*, Black Dog and Leventhal Publishers, New York, NY., c2004.

Smith, Gar, *Nuclear Roulette*, Chelsea Green Publishing, White River Junction, Vermont, c2012.

Some, Sobonfu E., *Welcome Spirit Home*, New World Library, Novato, CA, c1999

Stiglitz, Joseph E., *The Price of Inequality*, W. W. Norton and Company, c2012

Still, William, *The Underground Railroad*, Ebony Classics, written in 1871. Reprinted by Johnson Publishing Company, Chicago, ©1970.

Swan, Michael, *Doctrine of Discovery First Repudiated in 1537*, writing for The Catholic Register, October 2, 2014

Tanaki, Ronald, *A Larger History*: A History of our Diversity, with Voices, Little, Brown and Company, ©1998.

Tarbuck, Edward J. and Lutgens, Frederick K., *Earth Science*, 11th edition, Pearson Prentice Hall, c2006.

Tarbuck, Edward J. and Lutgens, Frederick K., *Earth Science, California edition*, Pearson Prentice Hall, c2006.

Trinidad and Tobago, Government of: *National Report on Human Settlement*, presented to Habitat: the United Nations Conference on Human Settlement, Vancouver, May 31 to June 11, 1976.

Trinidad and Tobago, Government of: *Five-Year Economic Programme, 1956-1960*, Memorandum for 1957, Approved by the Legislative Council, 21 December 1957.

Trinidad and Tobago, Government of: *Five Year Development Programme, 1958 – 1962*, Planning and Development Unit, Port-of Spain, Trinidad, 28 December 1957.

Trinidad and Tobago, Government of: *Draft Second Five-Year Plan, 1964-1968*, National Planning Commission, ©1963.

Trinidad and Tobago, Government of: *Report of the Economic Commission*, appointed to examine Proposals for Association within the Framework of a Unitary State of Grenada and Trinidad and Tobago, January 1965.

Trinidad and Tobago, Government of: *Grenada Five-Year Development Plan, 1964-1968*, Report of the Development Programme Commission, ©1965.

Trinidad and Tobago, Government of: *Agreement Establishing the Caribbean Free Trade Association*, Principal and Supplementary Agreements, ©1968

Trinidad and Tobago, Government of: *Draft Third Five-Year Plan, 1969-1973*, Office of the Prime Minister, Eric Williams, 16 December 1968.

Udell, Emilie,*Unequal Risk*, A chronicle of studies, outlining the inadequacies of prevailing regulations to protect workers from work-related hazards. Released by The Center for Public Integrity, June 29, 2015.

United Nations, *The State of the World Environment, 1978*, by Dr. Mostafa Kamal Tolba, Executive Director of the United Nations Environment Programme, June 5, 1978.

United Stated, Department of Energy: *Residential Conversation Service Auditor Training Manual*, prepared by U. S. Department of Energy, with support from: Oak Ridge National Laboratory, University of Massachusetts Cooperative Extension Service Energy Education center and the Sloar Energy Research Institute, February 1, 1981.

United States, Department of Housing and Urban Development: *Neighborhoods:* A Self-help Sampler, Office of Neighborhoods, Voluntary Associations and Consumer Protection, October 1979.

United States, Department of Housing and Urban Development: *4th Biennial HUD Awards for Design Excellence*, October 1970.

United States, Department of Housing and Urban Development: *Development Programs*, August 1980.

United States, Department of Housing and Urban Development: *Kalamazoo Breakthrough Housing Venture*, National Corporation for Housing Partnerships, August 1971.

Upton, John, *Pulp Fiction,* The European Accounting Error that is Warming the Planet. ©2015 Climate Central. All rights reserved.

Vesper Society Group / Association of German Protestant Academies: *A Developing World Economy*, the Ethics of International Cooperation, presented in Conference at Santa Sabina Center, San Rafael, California, September 28 to October 1, 1989.

Walker, J. Samuel, *Three Mile Island,* A Nuclear Crisis in Historical Perspective, University of California Press, Berkeley, CA c2004

Wallerstein, I., *The Modern World System*: Capitalist Agriculture and the Origins of the European World-Economy in the Sixteenth Century, San Diego Academic Press. ©1974.

Weart, Spencer, *The Discovery of Global Warming*, Harvard University Press, ©2003.

West, Cornel, *Restoring Hope*, Conversations on the Future of Black America, Beacon Press, Boston, Massachusetts, c1997

West, Cornel, *The Cornel West Reader*, Basic Civitas Books, ©1999.

Westmacott, Richard, *African American Gardens and Yards in the Rural South*, The University of Tennessee Press, Knoxville, ©1992

Whitehead, John W., *A Government of Wolves*: The Emerging American Police State, the Rutherford Institute, Charlotte, Virginia, ©2015

Williams, Juan, *Eyes on the Prize*, America's Civil Rights Years, 1954-1965. Viking, New York, ©Blackside, Inc., 1987

Wink, Walter, *Engaging the Powers*: Discernment and Resistance in a World of Domination, Fortress Press, Minneapolis, c1992.

Woodmencey, Jim, *Weather in the Southwest*, Southwest Park and Monument Association, Tucson, Arizona, c2001

WorkSafe NB: *Health and Safety Orientation Guide for Employers*, prepared by WorkSafe, October 2011, Revised June 2014 www.worksafenb.ca/docs/WorkSafeNBOrientationGuide_e.pdf -

World Bank Group: *Policies and Operations*, September 1974.

World Bank Group: *Urbanization*, Sector Working Paper, June 1972.

World Bank Group: *Agricultural Credit*, Sector Policy Paper, May 1975.

World Bank Group: *Agricultural Land Settlement*, Issues Paper, January 1978.

World Bank Group: *Development Finance Companies*, Sector Policy Paper, April 1976.

World Bank Group: *Education*, Sector Working Paper, December 1974.

World Bank Group: *Employment and Development of Small Enterprises*, Sector Policy Paper, February 1978.

World Bank Group: *Environment and Development*, June 1975.

World Bank Group: *Health*, Sector Policy Paper, March 1975.

World Bank Group: *Housing*, Sector Policy Paper, May 1975.

World Bank Group: *Land Reform*, Sector Policy Paper, May 1975.

World Bank Group: *Rural Electrification*, Sector Policy Paper, October 1975.

World Bank Group: *Rural Development,* Sector Policy Paper, February 1975.

World Bank Group: *Rural Enterprise and Nonfarm Employment*, Sector Policy Paper, January 1978.

World Bank Group: *Village Water Supply*, Sector Policy Paper, March 1976.

Yen, Duen Hai, *Domination Systems*, Copyright © 2006 by Duen Hai Yen, All rights reserved. Last updated 15 February 2007

Zinn, Howard, *Original Zinn, Conversations on History and Politics*, Harper Perennial, New York, NY., c2006

Articles, Documentaries, Papers, Reports, Studies and other publications represented in this work

Abaffy, Luke, "Foundation Flaws Make Kentucky's Wolf Creek Dam a High-Risk Priority", February 27, 2012.

Abbott, Jeff, "Transnational Companies Driving Deadly Conflict in Guatemalan Indigenous Territory", Truthout Report, 15 April 2015

Agre, Philip E., "What Is Conservatism and What Is Wrong with It?" August 2004, http://polaris.gseis.ucla.edu/pagre/conservatism.html

Alexander, Michelle, "Locked Out of America", in interview with Bill Moyers, Moyers & Company, December 20, 2013.

Alperovitz. Gar, "America Beyond Capitalism: An 'Evolutionary Reconstruction' of the System Is Necessary and Possible", A review of his book: *America Beyond Capitalism*, by Gar Alperovitz, Democracy Collaborative Press, 12 July 2012.

Alperovitz. Gar, "Inequality's Dead End — And the Possibility of a New, Long-Term Direction", Nonprofit Quarterly, Spring 2015, 17 March 2015.

Altawell, Najib, "Introduction to India's Energy and Proposed Rural Solar-PV Electrification". E-mail: n.altawell@dundee.ac.uk University of Dundee, Carnegie Building, Dundee DD1 4HN, Scotland, UK.

Amadeo, Kimberly, "What Makes Derivatives So Dangerous?", About News US Economy Expert, May 19, 2015.

Amadeo, Kimberly, "Where Were the 223,000 Jobs Added in April?" United States Economy Expert, Current U.S. Employment Statistics: Job Growth and Loss, May 8, 2015.

American Civil Liberty Union, "Racial Profiling", ACLU Report, July 18, 2014.

Andersen, Anne Vestergaard, "Clean Air Tours Launched in Cooperation with Danish Ministry of the Environment" State of Green, Denmark, 21 May 2015.

Andersen, Anne Vestergaard, "Record High Exports of Danish Energy Technology", 30 April 2015.

Anderson, Carol, "Ferguson isn't about black rage against cops. It's (about) White Rage Against Progress", 29 August 2014.

Anderson, Carol, "It's (about) White Rage Against Progress." African American Studies, (Ferguson isn't about black rage against cops.) Emory University, 29 August 2014.

Anthony, Carl, et. Al. "People's Tribunal on the Rights of Nature", by Carl Anthony, Brian Swimme, Anuradha Mittal, Courtney Cummings and Bill Twist, presenting "The Forum" at Laney College, October 5, 2014.

Appalachian Voices, "Learn More about Mountaintop Removal Coal Mining" by Appalachian Voices, Boone, NC c2006.

Appalachian Voices, "Impacts of Mountaintop Removal: presents a series of studies and reports on a variety of impacts". © 2013 Appalachian Voices, Boone, NC 28607

Appleby, Terry, "The Industrial Revolution Gave Birth to the Cooperative Movement", Wedge Community Newsletter, October-November 1998.

Associated Press, "Colorado River begins flooding Mexican delta", March 27, 2014

Associated Press, "Flash Flood Warnings Extended as Much-Needed Rainstorm is Expected to Bring More Showers to Parched Southern California" 3 December 2014.

Associated Press, "French supermarkets told to give away unsold food to charity or farms", for The Telegraph, 22 May 2015.

Associated Press, "New San Francisco-Oakland Bay Bridge opens at last to replace the one damaged in 1989 earthquake after 12 YEARS of construction," 3 September 2013.

Associated Press, "President Pressures Lawmakers to Back $302 billion Transportation Plan and Fix Nation's Decaying Roads and Bridges", 14 May 2014 –

Associated Press, Report: Warming water in LI Sound altering fish populations, June 8, 2015.

Atkin, Emily, "A Huge Antarctic Ice Sheet Is Melting Three Times Faster Than Previously Thought", Climate progress, December 3, 2014.

Atwood, Margaret, "It's not Climate Change; It's Everything Change," Animations by Carl Burton, July 27, 2015.

Auch, Roger, et. al. "Urban Growth in American Cities: Glimpses of U.S. Urbanization by Roger Auch, Janis Taylor, and William Acevedo, January 2004

BajaInsider.com "A Historical Look at Hurricanes in Baja", published by online magazine, October 2013.

Baldera, Alexis, "Where Did the BP Deepwater Horizon Oil Go?" Exploring the 'bathtub ring' left from the explosion four years earlier, by October 31, 2014.

Barber, Gregory, "Are Cage-Free Eggs all They're Cracked Up to Be?" by Mother Jones, February 10, 2016.

Barron-Lopez, Laura, "Conservatives urge states to 'fiercely resist' EPA's climate rule," December 5, 2014

Barry, Ellen, "Earthquake Devastates Nepal, Killing More Than 1,900", New York Times, April 25, 2015.

Bauerlein, Monika and Jeffery, Clara, "This is What's Missing from Journalism Right Now," Mother Jones Magazine, August 17, 2016.

Beck, Erin, "Thousands of gallons of oily-water mixture recovered from trenches along Kanawha River", staff writer, The Charleston Gazette, February 23, 2015.

Bell, Beverly, "Haitian Farmers Commit to Burning Monsanto Hybrid Seeds", Truthout Report, 18 May 2010.

Berman Jillian, "Big Banks Don't Pay A Third Of Tellers Enough To Live On": The Hiffington Post. 23 January 2014

Bernd, Candice, "Nuclear Reactors Are Toxic to Surrounding Areas, Especially With Age". Energy - Nuclear Plant Study, Truthout Report, 11 March 2014.

Bernd, Candice, "Police Departments Ignore Rampant Sexual Assault by Officers", Truthout News Analysis, July 2, 2014.

Bernd, Candice, "Police Departments Retaliate Against Organized "Cop Watch" Groups Across the US", Truthout Report, 2 October 2014.

Bernd, Candice, "Why There's a Real Chance My Texas Town Might Ban Fracking", Truthout News Analysis, 24 October 2014

Bidwell, Allie, "Early Childhood Poverty Damages Brain Development, Study Finds", US News and World Report, October 28, 2013.

Biello, David, "Spent Nuclear Fuel: A Trash Heap Deadly for 250,000 Years or a Renewable Energy Source?" Scientific American, 28 January 2009.

Biello, David, "Ten possibilities for staving off catastrophic climate change", National Aeronautics and Space Administration, November 26, 2007.

Bienkowski, Brian, "A Sacred Water Story" Wisconsin's Menominee Native Cultural Rebirth Underway in North Dakota, Environmental Health News, September 29, 2016.

Bienkowski, Brian, "Hydraulic fracturing wells and the pollution from them are more likely to impact poor communities in Pennsylvania", Environmental Health News, 6 May 2015.

Bigelow, Bill "Clean Coal Propaganda in the Schools", Yes! Magazine, Op-Ed, 8 March 2014.

Bilefsky, Dan, "David Cameron Grapples with Issues of Slavery Reparations in Jamaica", for New York Times, October 1, 2015.

Birckhead, Tamar, "Delinquent by Reason of Poverty", August 20, 2012

Biswas, Asit and Kirchherr, Julian, "Fracking, Shale Gas, Keystone and Water Contamination", Huffington Post, 23 September 2013.

Biswas, Asit and Kirchherr, Julian, "Shale Gas: Black Hole for Water", May 29, 2913.

Bittman, Mark, "Now This Is Natural Food", New York Times, October 22, 2013. - NYTimes.com

Blake, Mariah, "How Hillary Clinton's State Department Sold Fracking To The World", Mother Jones, September 10, 2014.

Blodget, Henru, "Profits Just Hit Another All-Time High, Wages Just Hit Another All-Time Low", Post-Recession Trends, April 11, 2013.

Boehmer-Christiansen, Dr Sonja, editor: Energy and Environment, www.multi-science.co.uk a journal of Science and Technology, Volume 26, 2015

Borosage, Robert L., "Let's Give Workers Something to Cheer About", by OtherWords, Op-Ed., 1 September 2014.

Boston.com "Dust storm in Australia - The Big Picture", September 23, 2009.

Bouie, Jamelle, "The Most Brazen Attempt at Voter Suppression Yet", Slate staff writer, November 2014.

Brain, Marshall, "How Sewer and Septic Systems Work", an illustrated description of septic systems. November 7, 2006.

Braun, David, "Skull in Underwater Cave May Be Earliest Trace of First Americans", News Watch, National Geographic Society, February 18, 2011.

Breen, Cathy, "Reflecting on My Visits to Families in Exhausted Iraq", Voice of Creative Nonviolence, Op-Ed., 23 November 2012.

Breyman, Steve, "Why Good News for the Ozone Is Bad News for the Climate", Truthout News Analysis, 27 September 2014.

British Broadcasting Corporation, "Migrant Crisis: Migration to Europe Explained in Graphics", a British Broadcasting Corporation report, 27 October 2015.

British Library Board, "The Fight for Universal Suffrage", Published by the British Library Board, ©1936.

Brooks, David, "The Power of a Dinner Table," by David Brooks, for New York Times, October 18, 2016.

Brown, Ellen, "Seniors' Social Security Garnished for Student Debts", Truthout I News Analysis, 11 May 2012

Brown, Paul, "Beyond Nuclear" Nuclear Subsidy Deal 'Will Kill Renewables' - Climate News Network, ©2009 Beyond Nuclear.

Brown, Paul, "Nuclear Subsidy Deal 'Will Kill Renewables'", Climate News Network, April 7, 2014.

Bryan, Wright, "Iraq WMD Timeline: How the Mystery Unraveled," compiled and produced with additional research by Douglas Hopper from a variety of United Nations documents, November 15, 2005.

Buchanan, Larry, "Your Contribution to the California Drought", by Larry Buchanan, Josh Keller, and Haeyoun Park Photos and video by Tony Cenicola and Dave Frank, May 4,2015, ©2015, The New York Times

Butler, Rhett, "The Amazon: The World's Largest Rainforest", ©2014, at: http://rainforests.mongabay.com/amazon/#sthash.JSQuDehn.dpuf

Butler, Rhett, "Where Rainforests are Located: Biogeographic Tropical Forest Realms", July 22, 2007.

Buntin Simmons B., "Origins and History of Suisun City's Downtown Redevelopment," Originally appeared in Issue No. 3. © Copyright

1997-2013 *Terrain.org: A Journal of the Built + Natural Environments.* All rights reserved.

Burghardt, Tom, "Narcotics: CIA-Pentagon Death Squads and Mexico's 'War on Drugs'", Global Research, May 28, 2012. Copyright © Tom Burghardt, Antifascist Calling..., 2012.

Burnett, Jasmine, "What Black Women in West Philadelphia Had to Say About Women's Equality Day", RH Reality Check, News Analysis, 1 September 2014

Burns, Mike, Groch-Begley, Hannah, Shere, David, Tone, Hilary, & Uwimana, Solange, "10 Myths Conservative Media Will Use Against Immigration Reform", researched February 1, 2013

Butler, Kiera, "One Disease Hits Mostly People of Color. One Mostly Whites. Which One Gets Billions In Funding?" Mother Jones, 4 May 2015.

Butrica, Barbara A., Iams, Howard M., and Smith, Karen E., "The Changing Impact of Social Security on Retirement Income in the United States", Social Security Bulletin, Vol. 65 No. 3, 2003-2004.

Byrd, Deborah, "Huge New Antarctic Marine Reserve", EarthSky Update, Earth-Human World, October 28, 2016.

Canon, Gabrielle, "17 Everyday Items That Use a Whole Lot of Water", April 21, 2015

Cantu, Aaron, "How the Mainstream Media Helped Kill Michael Brown", Truthout, News Analysis, 12 August 2014.

Cantu, Aaron, "Ring of Snitches: How Detroit Police Slapped False Murder Convictions on Young Black Men", Truthout, 31 March 2015

Cantu, Aaron Miguel, "New Report: Fortune 100 Companies Have Received a Whopping $1.2 Trillion in Corporate Welfare Recently", (Military contractors, oil companies and banks are the biggest 'welfare queens' around). AlterNet Report, 20 March 2014.

Cape Town City Hall, "Non-European United Front of South Africa: Minutes of Conference held in City Hall," Cape Town, 8-10 April 1939.

Cardiff, Dennis, "Gotta Find a Home", Conversations with Street People, Ottawa Innercity Ministries, April 2013.

Carey, Bjorn, "Sahara Desert Was Once Lush and Populated", July 20, 2006.

"Salvadoran Farmers Successfully Oppose the Use of Monsanto Seeds", by Dahr Jamail, Truthout, 8 July 2014

Carr, Jane Greenway, "Chipping at the Polish": The Story Behind Sarah Maslin Nir's Nail Salon Investigation, June 17, 2015.

Cart, Julie, "Central Valley's growing concern: Crops raised with oil field water", Los Angeles Times, ©2014.

Carus, Felicity, "Dangerous Levels of Radioactivity Found at a Fracking Waste Site in Pennsylvania: A new study found elevated levels of chloride, bromide, and other chemicals in Marcellus Shale wastewater," MotherJones Climate Desk, October 3, 2013.

Castaneda, Jorge G. and Massey, Douglas S., "Do-It-Yourself Immigration Reform," June 1, 2012. Jorge G. Castaneda the foreign minister of Mexico from 2000 to 2003, Douglas S. Massey is a professor of politics and Latin American and Caribbean Studies at New York University and a professor of sociology and public affairs at Princeton.

Castle, Stephen and Rudore, Jodi, "A Symbolic Vote in Britain Recognizes a Palestinian State", Written for the New York Times, October 13, 2014.

Cawthorne, Alexandra, "The Straight Facts on Women in Poverty", Center for American Progress, October 8, 2008.

Cazenave, Noel, "Understanding Our Many Fergusons: Kill Lines - the Will, the Right and the Need to Kill", Truthout Op Ed., 29 September 2014.

Center for Climate and Energy Solutions: "Leveraging Natural Gas to Reduce Greenhouse Gas Emissions, June 2013.

Center for Food Safety, "True Food Shoppers Guide to Avoiding GE Food", Publications & Resources, January 01, 2014.

Chang, Chih-Hung, "Energy Breakthrough Uses Sun to Create Solar Energy Materials", Professor of Chemical Engineering, Oregon State University, April 3, 2014.

Chaykin, Sasha, "As kidney disease kills thousands across continents, scientists scramble for answers", September 17, 2012.

Chen, Michelle, "About 620,000 Military Families Rely on Food Pantries to Meet Basic Needs", published by The Nation, Report, ©2014.

Chicago Maroon, "Community Gardens", a Research Paper, prepared by the Maroon Editorial Board of the Chicago Maroon, Student Newspaper, University of Chicago, 15 May 2012.

Childress, Sarah, "Why Voter ID Laws Aren't Really about Fraud", October 20, 2014.

Choma, Russ, "Green Jobs: A Promise Unfulfilled," Investigative Reporting Workshop, News Report, 8 September 2011

Chomsky, Aviva, "America's Continuing Border Crisis: The Real Story Behind the 'Invasion' of the Children", August 26, 2014.

Chomsky, Aviva, "The United States' Continuing Border Crisis: The Real Story Behind the 'Invasion' of the Children", TomDispatch, News Analysis, 25 August 2014.

Christie, Les, "Uninsured in America: Number of people without health insurance in U.S. climbs" @CNNMoney September 13, 2011.

City University of New York, "Nuclear Waste". A documentary published by City University of New York, 2011.

Clayton, Lawrence, "Bartolome de las Casas: A Brief Outline of his Life and Labor," Excerpts included from the works of David Orique, O.P., Biography, June 29,2012.

CoalSwarm, "Mine Safety: A manual promoting the health and safety of miners on the job". Prepared by CoalSwarm and the Center for Media and Democracy, 2011.

Codevilla, Angelo M., "America's Ruling Class — And the Perils of Revolution" from The American Spectator, issue July-August 2010.

Cole, Nicki Lisa, "CO2 Emissions Growth Takes a Bite Out of Apple's Sustainability Claims" by Truthout, 22 April 2015.

Collier, Marty, "Fair Trade Helps Farmers Prosper", copyright 2001 New World Outlook, nwo@gbgm-umc.org.

Collier, Victoria, "Computerized Vote Rigging Is Still the Unseen Threat to US Democracy: It's Time to Change the System", Truthout, News Analysis, 3 November 2014.

Collier, Victoria and Cummins, Ronnie, "Occupy Rigged Elections: A Call for the Second American Revolution in 2012", Truthout Op-Ed, 27 December 2011.

Collins, Chuck, "Beyond Divestment: Climate-Concerned Philanthropists Pledge to Move Billions to Wind and Solar", YES! Magazine, Report, 18 September 2014.

Collins, Mary Ann, "Were the Early Christians Roman Catholics?", ©February 2002 www.CatholicConcerns.com

Conner, Phillip, Cohn, D'Vera and Gonzalea-Barrera, Ana, "Changing Patterns of Global Migration and Remittances," for Pew Research Center, Washington D.C. 20036, December 17, 2013.

Conniff Richard, "The Myth of Clean Coal," in Business & Innovation, Climate, Energy, Science & Technology, Sustainability, North America. 3 June 2008.

Consumer Education, Teaching and Learning for a Sustainable Future, Module 9: ©UNESCO 2010.

Cook, Roberta L.., "Fundamental Forces Affecting U.S. Fresh Produce Growers and Marketers", (cook@primal.ucdavis.edu), Cooperative Extension Specialist, University of California, Davis.

Costa Rica Tourism Board, "Tips for Having a Great and Safe Trip", by the Costa Rica Tourism Board, San Jose, Costa Rica.

Crater Lakes, "12 Most Incredible Crater Lakes On Earth: Photography and text by scribol.com/environment/12-most-incredible-crater-lakes-on-earth

Crawford, Amy, "What Traditional Societies Can Teach You About Life", commenting on Jared Diamond's *The World Until Yesterday*: What Can We Learn from Traditional Societies? December 26, 2012. Smithsonian.com

Crossa, Mateo, "U.S. Ties with Mexico's Military Have Never Been Closer": An Interview with Jesse Franzblau, for the Center for Economic and Policy Research, (CEPR), 20 May 2015.

Cuffe, Sandra, "Militarization of the Mesoamerican Barrier Reef Harms Indigenous Communities", Truthout News Analysis, 17 May 2014.

Currier, Cora, "The 24 States That Have Sweeping Self-Defense Laws Just Like Florida's" ProPublica, March 22, 2012.

D'Almeida, Kanya, "In U. S. Prisons Psychiatric Disability is Often Met with Brute Force", Truthout Report, 18 July 2015.

Daley, David, "How the GOP's Cynical Election Strategy is Imploding". YES Magazine, August 3, 2016.

Danish Energy Agency, "New Model for Calculating the Economy of Energy Efficiency", October 19, 2016.

Dannin, Ellen, "Big Players Promote Water Privatization", Truthout, News Analysis, 1 Innovations or Hucksterism? August 2014.

Dannin, Ellen, Three Little-Known Infrastructure Privatization Problems, Truthout, News Analysis, 22 December 2014

Darder, Antonia, "Countering the 'Chega Freire' Campaign in Brazil: Our Emancipatory Struggle Continues", Truthout, 22 April 2015.

Davenport, Coral and Robertson, Campbell, "Resettling the First American, [US] 'Climate Refugees'," May 2, 2016.

Del Savio, Lorenzo and Mameli, Matteo, "The Evaporation of Democracy", Truthout, 20 October 2014.

Derber, Charles and Magrass, Yale, "Romney's '47 Percent' Blunder Reveals the Hidden Heart of His Agenda," Truthout, 09 October 2012 |

Derber, Charles and Magrass, Yale, "When Wars Come Home", Truthout Op-Ed, 19 February 2013.

DeSilver, Drew, "Who makes minimum wage?" Pew Research Center, Sept. 8, 2014

Dexing, Qin, "Ruling set to burn US solar firms," China Daily Web Editor, 25 December 2014

Deppen, Colin, "DEP Report Details Legacy of Coal Mining", ERA Report, January 1, 2015.

Desert Sun, "Little Oversight as Nestle Taps Morongo Reservation Water," (with a permit that expired in 1988), The Desert Sun, July 12, 2014.

Devereux, Charlie, "Argentine Flights Axed as Chile Volcano Ash Reaches Buenos Aires", April 24, 2015.

Devlin, Hannah, "Early Men and Women were Equal," Says Scientists, Science Correspondent, @hannahdev 15 May 2015.

Deyle, Robert E., Bailey, Katherine C. and Matheny, Anthony, "Adaptive Response Planning to Sea Level Rise in Florida and Implications for Comprehensive and Public-Facilities Planning" by Florida Planning and Development Lab, Department of Urban and Regional Planning, Florida State University, September 1, 2007

Di Leo, Jeffery R., et.al. "Twelve Theses on Education's Future in the Age of Neoliberalism and Terrorism", by Jeffery R. Di Leo, Henry Giroux, Kenneth J. Saltman, and Sophia A. MvcClennen, Paradigm Publishers, Book Excerpt, 4 September 2014.

Dieffenbacher-Krall, John, "Rejecting the Christian Doctrine of Discovery and Forging a New Relationship with North America's Indigenous People": A Sermon Preached by The Rev. John Dieffenbacher-Krall, at St. David's Episcopal Church, Barneveld, NY in the Central Diocese of NY. July 12, 2009.

Dillivan, K. D., Balko, Radley, "Where Does the Money Go? Food Marketing Margins Explained", "How municipalities in St. Louis County, Mo., profit from poverty", 3 September 2014. Follow @ radleybalko

Dobbs, Michael, "Evidence of a Serious Conflict of Interests in the U.S. Government over Iraq", Washington Post, 30 Dec 2002.

Dodd, John, "Unemployed Americans Speak Out as Benefits are Slashed at Christmas", Truthout Opinion, Friday, 14 February 2014 / truth-out.org

Dollman, Darla, "Dust Storms and Haboobs: Dangerous Consequence of Drought", August 12, 2013

Dolinar, Brian and Kilgore, James, "Carceral Conglomerate" Makes Millions From Incarcerated, Their Friends and Families, Truthout News Analysis, 13 February 2015.

Douglas, Bruce, "Anger Rises as Brazilian Mine Disaster Threatens River and Sea with Toxic Mud," by Bruce Douglas in Vila de Regencia, for the Guardian, 22 November 2015.

Drouin, Roger, "Contaminated Water Supplies, Health Concerns Accumulate With Fracking Boom in Pennsylvania", Truthout, News Analysis, 14 March 2014.

Duen Hai Yen, "Domination Systems" by Duen Hai Yen, Copyright ©2006 All rights reserved. Last updated 15 February 2007

Dunlap, David W., "Crews at Work Replacing Section of Delaware Aqueduct", New York Times, November 19, 2014

Dupere, Katie, "7 important ocean trends, and what we can do about them", National Oceanic and Atmospheric Administration, June 8, 2015.

Dvorak, Petula, "A Bittersweet Farewell to the Obamas at their Last White House Easter Egg Roll", by Local Columnist, Washington D.C., March 28, 2016.

Dvorak, Petula, "Background Report: Donald Trump: Mainstreaming a Hate Movement", Washington Post, August 23, 2016.

Dvorak, Petula, "The Obamas: Refusing to Give in to the Haters after Years of Threats and Abuse," Local Columnist, Washington D.C., March 31, 2016.

Dvorak, Petula, "The 'Trump Effect' is Contaminating Our Kids – and could Resonate for Years," Local Columnist, Washington D.C., March 7, 2016.

Dykstra, Peter, "Opinion: Doubling down on doubt, (Climate Change Opposition"). Environmental Health News/The Daily Climate, March 5, 2015

ENENews, "Radioactive contamination was reported in the city of Carlsbad, New Mexico, 25 miles away from leakage discovered at the U.S. #WIPP nuclear waste storage facility. The leaking was detected on Saturday, but was not announced until Monday. The incident was published in a report a week later, February 24, 2014.

Eckert, Vera and Chestney, Nina, "Renewable energy surge revives Europe's power trade", Reuters/ Heinz-Peter Bader, December 22, 2014.

Editors of Encyclopedia Britannica, "Sandinista", by The Editors of Encyclopedia Britannica, ©2013.

Edwards, Nick, "The Andaman Islands: India's far flung string of pearls", February 3rd, 2015

Egan, Timothy, "A Mudslide, Foretold", Contributing Op-Ed Writer - NYTimes.com SundayReview – March 29, 2014.

Ehrenreich Barbara, "On Turning Poverty Into an American Crime," TomDispatch | Truthout Op-Ed., Tuesday 9 August 2011.

Eisen, Lauren-Brooke, "Charging Inmates Perpetuates Mass Incarceration", Brennan Center for Justice, May 21, 2015.

Eligon, John, "Crackdown in a Detroit Stripped of Metal Parts", New York Times, 15 March 2015

Emanuele, Vincent, "Christian Parenti on Climate Change, Militarism, Neoliberalism and the State", Truthout, Interview, 12 May 2015.

Emerge Magazine, "Crime Pays", a documentary: 'Cashing in on Black Prisoners'", October 1997.

Ernst, Aaron, "Why New Yorkers should be worried about their water supply", @aaronernst and Christof Putzel @ChristofPutzel, Aljazeera, America Tonight, August 4, 2014

Ernsting, Almuth, "Abundant Clean Renewables? Think Again!" Truthout, News Analysis, 16 November 2014.

Eternity Max, "The 'Urbee' 3D-Printed Car: Coast to Coast on 10 Gallons?", Truthout interviews Jim Kor, designer of the car, 23 November 2014.

Farm Workers News, The Plight of Farm Workers, a report presented by Migrant Farm Workers Division, Colorado Legal Services.

Faught, Michael K., "Remote Sensing, Target Identification and Testing for Submerged Prehistoric Sites in Florida: Process and Protocol in Underwater CRM Projects". Panamerican Consultants, Inc., Tallahassee, Florida, mfaught@comcast.net

Ferdman, Roberto A. "Cholesterol in the Diet: The Long Slide from Public Menace to No 'Appreciable' Effect," for The Washington Post, February 19, 2015.

Ferdman, Roberto A. and Peter Whoriskey, "Think of Earth, not just your stomach, panel advises", for The Washington Post, February 19, 2015

Field, Tory and Bell, Beverly, "Food Justice: Connecting Farm to Community", Harvesting Justice Series, Truthout, 1 July 2013.

Finnegan, William. "Leasing the Rain, Bechtel vs Bolivia: Details of the Case and the Campaign," The New Yorker, April 2000.

Fischer, Douglas, "Dark Money: Funds Climate Change Denial Effort", The Daily Climate, December 23, 2013.

Fisher, Max, "Here's a Map of the Countries That Provide Universal Health Care", (America's Still Not on It), The Atlantic, June 28, 2012.

Fisher, Max, "What is Modern Slavery?" using data from the United States Department of State, February, 2013.

Flanders Laura, "The 1% Should Be Afraid: The New Norm in the Workplace Is Unstable", Truthout, Interview and Video, 11 February 2014.

Flowers, Margaret and Zeese, Kevin, "The Secret Rise of 21st Century Democracy", Truthout, 20 February 2013.

Follman, Mark, et.al., "What Does Gun Violence Really Cost?" By Mark Follman, Julia Lurie, Jaeah Lee and James West, Mother Jones, April 17, 2015.

Foner, Eric and Garraty, John A., Editors, "Manifest Destiny", excerpt from: The Reader's Companion to American History. Copyright © 1991 by Houghton Mifflin Harcourt Publishing Company. All rights reserved.

Food and Water Watch, "Selling Out Consumers: How Water Prices Increase After Privatization", Fact Sheet, June 2011.

Forbes, Dr. Greg, Severe Weather Expert, "2011's Tornado Superoutbreak" with Introduction by Senior Meteorologist Stu Ostro (Twitter). Video: The Tuscaloosa Tornado of 2011.

Fountain, Henry, "Panel Urges Research on Geoengineering as a Tool Against Climate Change", New York Times, February 10, 2015.

Fountain, Henry, "Researchers Aim to Put Carbon Dioxide Back to Work," May 2, 2016.

Franklin, Jonathan, "Chilean Miners Fight Their Demons," The Daily Beast.

Fraser Institute, "What are the water quality concerns at mines?" ©2012 Fraser Institute. All rights reserved. www.fraserinstitute.org - See more at: http://www.miningfacts.org/Environment/What-are-the-water-quality-concerns-at-mines-/#sthash.sDp3DKSC.dpuf

Freedman Andrew, "BP's chairman lays out company's views on global warming", April 16, 2015

Fretz, Eileen, Flood Policy Director, "Protecting Rivers & Your Clean Water", December 21, 2012 / A "River Blog: Floods & Floodplains", St. Cloud, MN flooding in 2011, Mississippi River Mayors Advise Communities Affected By Sandy, Dave Swarz.

Fuels America, "Why Renewable Fuel Matters", a commentary on other alternatives to gasoline as fuel to burn to produce power. © 2014 Fuels America

Gabriel, Trip, "West Virginia Coal Country Sees New Era as Donald Blankenship Is Indicted", November 30, 2014.

Gadoua, Renee K., "Nuns Blast Catholic Church's 'Doctrine Of Discovery' That Justified Indigenous Oppression", Religion News Service, Posted: 09/10/2014.

Game of Village "What is Village", prepared by The Game of Village, ©2007

Garcia, Carla, Elliott, Lauren and Bell, Beverly, "Not On Our Land: Land Recovery Kicks Off in Honduras", Other Worlds, Op-Ed, 1 September 2012.

Garofalo, Pat, "Homeless Veterans, By the Numbers," November 11, 2011

Garrett, Wilbur E., "La Ruta Maya," by editor, Journal of the National Geographic Society, volume 176, #4, October 1989.

Garrison, Jessica, "L.A. engineers are puzzled by uptick in water pipe failures", Los Angeles Times, September 16, 2009

Garthwaite, Josie, Pictures: "Ford Solar Car at CES 2014, Past Sun-Power Vehicles", January 9, 2014.

Gelles, David. "How Producing Clean Power Turned Out to be a Messy Business". 13 August 2016.

Gertz, Emily J., Assoc. Editor, "Rivers Flowing into the Great Lakes are Teeming with Microplastic Pollution", Environment and Wildlife, TakePart, September 17, 2016.

Gilder Lehman Collection, "The Doctrine of Discovery, 1493." Pope Alexander VI's Demarcation Bull, May 4, 1493. (Gilder Lehman Collection). A primary source by Pope Alexander VI.

Gillis, Justin, "For Faithful, Social Justice Goals Demand Action on Environment", June 20, 2015.

Gillis, Justin, "Looks Like Rain Again. And Again", Commentary on global warming and increased rainfall, May 12, 2014.

Gillam, Carey, "Monsanto seeks retraction for report linking herbicide to cancer", Reuters, 24 March 2015.

Giroux, Henry A., "America's Addiction to Torture", Truthout | News Analysis, 17 December 2014.

Giroux, Henry A., "Dangerous Pedagogy in the Age of Casino Capitalism and Religious Fundamentalism", Truthout, News Analysis, 30 March 2012.

Giroux, Henry A., "Domestic Terrorism, Youth and the Politics of Disposability", by Henry A. Giroux, 21 April 2015.

Giroux, Henry A., "Left Behind? American Youth and the Global Fight for Democracy", truthout, Op-Ed., 28 February 2011.

Giroux, Henry A., "Murder, Incorporated: Guns and the growing culture of Violence in the US", Truthout News Analysis, 7 October 2015.

Giroux, Henry A., "The Powell Memo and the Teaching Machines of Right-wing Extremists", Truthout, 1 October 2009.

Giroux, Henry A., "The Racist Killing Fields in the US: The Death of Sandra Bland", Truthout News Analysis, 19 July 2015.

Giroux, Henry A., "The 'Suicidal State' and the War on Youth", by Henry A Giroux, Truthout Op-Ed, 10 April 2012.

Giroux, Henry A., "The United States' War on Youth: From Schools to Debtors' Prisons," Truthout News Analysis, 19 October 2016.

Gladstone, Rick, "Many Ask, Why Not Call Church Shooting Terrorism?" Published by Reuters, June 18, 2015.

Glick, Daniel, "Cinderella Story: Chattanooga Transformed", National Wildlife Magazine Feb/Mar 1996

Glionna, John, "Drought -- and neighbors -- press Las Vegas to conserve water", Los Angeles Times, © 2014

Goffman, Ethan, "Environmental Impacts of Wind Power", a study prepared by the Union of Concerned Scientists, Science for a Healthier and Safer World, Also: Capturing the Wind: Power for the 21st Century, June 2008.

Goldenberg, Suzanne, "Secret Funding Helped Build Vast Network of Climate Denial Think tanks," U.S. Environmental Correspondent to the Guardian, 14 February 2013.

Goldenberg, Suzanne, "Methane Leaks May Negate Climate Benefits of Natural Gas," The Guardian, June 5, 2013.

Goldstein, Alexis, "Can a 'Firewall Strategy' Keep Big Energy Out of Climate Talks? It worked for Fighting Tobacco", YES! Magazine, News Analysis, 22 September 2014.

Goldstein, David Goldy, "$15 and Change: How Seattle Led the Country's Wage Revolution", YES! Magazine, 31 August 2014.

Goldwag, Arthur, "The Measure of a Nation" Challenges Illusions of American Superiority, Truthout, 07 October 2012.

Gonzalez-Barrera, Ana, Krogstad, Jens Manuel and Lopez, Mark Hugo, "DHS: Violence, poverty, is driving children to flee Central America to U.S.", Pew Research Center, July 1, 2014

Goode, Erica, "Farmers Put Down the Plow for More Productive Soil", New York Times, March 9, 2015.

Goodman, Amy and Gonzalez, Juan, "Dog Whistle Politics: How Politicians Use Coded Racism to Push Through Policies Hurting All". Democracy Now, News Hour, Tuesday January 14, 2014.

Goodman, Amy and Gonzalez, Juan, "Heralding a National Trend? Enrollment Surges as New York City Begins Full-Day Prekindergarten for 50,000 Kids", Democracy Now! 4 September 2014.

Google, "Lake Nyos Story," reported by Google. See www.google.com Enter: Lake Nyos Story.

Gordon, Rebecca, "America's War on Drugs is Depressingly Similar to the Global War on Terrorism", written for Mother Jones, March 25, 2015.

Gratz, Roberta Brandes, "After Katrina, the Residents of New Orleans Saved Themselves", Nation Books, Book Excerpt, 12 June 2015.

Grillo, Iona, "U.S. Troups Increase Aid to Mexico in Drug War" from GlobalPost, October 6, 2011.

Gross, Benjamin, "Global Ocean Commission: Our Oceans Are in Decline", July 4, 2014

Gross-Galiano, Dr. Socorro, "Pan-American Alliance on Nutrition and Development for the Achievement of the MDGs", Pan American Health Organization Assistant Director, March 20, 2009.

Grossman, Elizabeth, "Five Years After the BP/Deepwater Horizon Disaster, Oil Spills Are on the Rise", Earth Island Journal, 19 April 2015.

Grossman, Elizabeth, "New UN Report Shows Just How Awful Globalization and Informal Employment are for Workers," In These Times, 28 October 2016.

Grossman, Elizabeth, "Study Links Widely Used Pesticides to Antibiotic Resistance" 24 March 2015.

Guerra, Crystal Vance, "The Dirty War Against Youth, From Ferguson to Ayotzinapa", Truthout News Analysis, 5 September 2015.

Guide to Visual and Other Clues Indicating the Risk of Sinkholes: Green links show where you are. © Copyright 2015 InspectApedia.com, All Rights Reserved.

Gundersen, Arnie, "Should Japan Restart Its Nuclear Reactors?" The Mark News, Op-Ed, 29 August 2014.

Hackman, Rose, "Belize plan to allow offshore drilling threatens Great Blue Hole, say critics", The Guardian, 10 May 2015.

Hager, L Michael and Roscoe, Lee, "Boston Endangered: Time To Close the Pilgrim Reactor", Truthout | Op-Ed, 23 March 2014 / TRUTH-OUT.ORG

Halberstadt, Jason, "Otavalo: World Famous Indigenous Market", All Rights Reserved, 7 May 2013

Halsall, Paul, "Feudalism: Charlemagne and the Holy Roman Empire, (Origin of the European Divine Rights of Kings), Education Portal, Modern Western Civilization", Copyright 2003-2014. www.fordham.edu

Halweil, Brian, "Organic Gold Rush". Brian Halweil, Research Associate, WorldWatch Institute. World Watch Magazine, May/June 2001, volume 14, #3.

Hamer, Dr. Mary M.D., "Apology to the Native American Indians," Countercurents.org, 8 December 2009.

Haneberg, Bill, "What can I do to Prepare for an Earthquake?" New Mexico Bureau of Geology and Mineral Services, Cartoons by Jan Thomas, 19 May 2016.

Hanley, Charles J., "The American 'allergy' to global warming: Why?", Associated Press Special Correspondent, September 23, 2011.

Hanlon, Mike, "World Population Density", ©2015 Global –Rural Mapping Project

Hansman, Heather, "How Ski Resorts Are Fighting Climate Change," February 26, 2015.

Hansen, Mary, "How This Library Paid $1 to Install Its Solar Panels", YES Magazine, June 2, 2015.

Hanson, J. "Child Labor in U.S. History," for the Child Labor Public Education Project: Sponsored by the University of Iowa Labor Center and The University of Iowa Center for Human Rights' Child Labor Research Initiative. ©Comcast Interactive Media 2016. Trickey, Erick, "How Hostile Poll-Watchers Could Hand Pennsylvania to Trump," October 2, 2016.

Hanson, Victor Davis, "California, an Engineered Drought", April 2, 2015.

Hare, Dr. Jonathan, "Solar Heaters and Other Parabolic Devices", The Creative Science Centre, Sussex University, ©2007.

Hargreaves, Steve, "How income inequality hurts America: It stifles economic growth", CNN, September 25, 2013.

Harris, Ayanna Banks, "The Forgotten Mothers: (The plight of mothers in prison)", Truthout Interview, 11 May 2014.

Harte, Alexis, "A City Guided by Its River", a story of San Antonia, Texas, American Forests, Summer 2003

Hartmann, Thom, "Are Economic Royalists Leading the US Over a Precipice?" Truthout, Twelve –Book Excerpt, 14 February 2014. Truth-out.org

Hartmann, Thom, "If Unions Are Breaking Automakers, Why Are BMW and Mercedes So Rich?", Yes! Magazine, News Analysis, 1 September 2014.

Hartmann, Thom, "Is This American Exceptionalism"? The Daily Take Team, Thom Hartmann Program, 14 October 2014.

Hauter, Wenonah, "Factory Farms Threaten Our Water, Air and Communities", Food and Water Watch, 27 May 2015. fwwatch.org

Haynes, H. Patricia, "Deadly Fukushima Crisis Further Corrodes Viability of Nuclear Energy", Truthout, Op-Ed., 11 March 2014.

Hays, Brooks, "NASA study: Net gains for Antarctic ice sheets", UPI, October 31, 2015.

Hays, Brooks, "Study: "Central Asia's Glaciers are Quickly Shrinking," from study published in Nature Geoscience, August 17, 2015.

Hays, Brooks, "Study details Greenland's Ice Sheet Plumbing System", for Nature, Published by UPI, October 9, 2015.

Hays, Brooks, "Study: Melting of Antarctica's Ice Shelves to Intensify" for Woods Hole Oceanographic Institute, Published bu UPI, October 12, 2015.

He, Ruoying and Weisberg, Robert, "Tides on the West Florida Shelf", College of Marine Science, University of South Florida, Journal of Physical Oceanography, October

Hearn, Kelly, "Toxic Coal in Tennessee": A massive coal sludge spill reveals the Tennessee Valley Authority has become a poster child for the failures of self-regulation, The Nation, February 4, 2009.

Hedges, Chris, "Let's Get This Class War Started", TruthDig, Op-Ed, 21 October 2013.

Hedges, Chris, "The Maimed Among Us", Truthdig, Op-Ed, 8 October 2012.

Hendee, David, "125 years ago today, Blizzard of 1888 ravaged the Plains", World-Herald Staff Writer, 12 January 2013.

Heikkinen, Niina, "A new price to pay for destruction of forests: a resurgence of the plague", Environment and Energy Daily Reporter, ClimateWire, 4 March 2015

Heikkinen, Niina, "Ocean's Oxygen Starts Running Low: Rising CO2 levels making it hard for fish to breathe," ClimateWire, May 2, 2016.

Henson Bob, "What are El Niño and La Niña?" National Center for Atmospheric Research | University Corporation for Atmospheric Research, Last updated: December 2014. @UCAR | http://www2.ucar.edu/news/backgrounders/el-nino-la-nina-enso

Hill David, "Will Colombia stop fighting coca by spraying glyphosate?" Guardian, 12 May 2015.

Hilts, P., "Haitians Held at Guantanamo Unconscious in a Hunger Strike." New York Times, 15 February 1993, p. A7.

History.com "Killer smog claims elderly victims" —This Day in History — 10/29/1948

Hodal, Kate, Kelly, Chris and Felicity, "Revealed: Asian slave labour producing prawns for supermarkets in US, UK", 10 June 2014.

Holland, Joshua, "America's Response to Child Refugees on the Border is Downright Shameful", July 11, 2014.

Holland, Joshua, "Land of the Free? US Has 25 Percent of the World's Prisoners", Presented on BillMoyers.com December 16, 2013.

Holthaus, Eric, "Biblical" Flooding Devastates South Carolina", Slate Climate Desk, October 5, 2015, www.slate.com

Hope, Jessica, 'The absolute right to rule' – The Divine Right of Kings, edited, 25 November 2014.

Hoppe, Robert, MacDonald, James, and Korb, Penni, "Small Farms in the United States: Persistence Under Pressure Economic Research Service", Economic Information Bulletin No. (EIB-63) 39 pp, February 2010.

Hornady, William T., "The Extermination of American Bison)" by William T. Hornady, (1889)

Howard, Brian Clark, "World's Largest Reserve Created off Antarctica," National Geographic, October 27, 2016.

Hudson, Adam, "Beyond Homan Square: US History Is Steeped in Torture" Truthout, 26 March 2015.

Hudson, Adam, "Early-Stage Gentrification: Richmond, California, Residents Push Back", Truthout Report, 17 July 2015.

Hudson, Drew, "Our Fight to Stop Water Privatization", Environmental Action, Washington D.C., 27 April 2014.

Huffington Post, "We Broke Iraq and We're Still Paying for the Damage", Posted: December 17, 2013.

Human Rights Watch. "Punishment and Prejudice: Racial Disparities in the War of Drugs", Human Rights Watch. New York. ©2000

Hunt, Tam, This Is Why It Makes Sense to Pair Solar With Electric Vehicles, 14 July 2014.

Hunter, John M., "Ascertaining Population Carrying Capacity under Traditional Systems of Agriculture in Developing Countries".

Reprinted from the Professional Geographer, University of Durham, volume 18, no. 3, May 1966.

Infoplease.com "Inquisition", http://www.infoplease.com/encyclopedia/society/inquisition- the-medieval-inquisition.html#ixzz3CGfx7N2J

International Labour Office, R183 – "Safety and Health in Mines Recommendation, 1995" (No. 183). Report on Meeting of the Governing Body of the International Labour Office, Geneva, Switzerland, 22 June 1995

Intensive Permaculture Design Course, "World's Largest Dam Removal Unleashes U.S. River After Century of Winter 2013, Intentional Permaculture Living in Hawaii by Lockerz, La'akea Community, February 2013

Jaffe, Sarah, "Released Occupy Activist Cecily McMillan: There's No Sense in Prison", Truthout Interview, 13 July 2014.

Jaffe, Sarah, "The Fight for Universal Pre-K: New York Charts a Checkered Path Toward Equal Early Education", Truthout News Analysis, 3 September 2014.

Jai, Shreya, "Government Chalks out Plans for Massive Solar Power Push", New Delhi, 28 October 2014.

Jamail, "Dahr, A Nation on the Brink: How US Policies Sealed Iraq's Fate", Truthout/TomDispatch, News Analysis, 17 July 2014.

Jamail, Dahr, "Addressing Population Growth - Through Freedom, Not Control - Is Crucial to Confronting Climate Disruption", Truthout, 22 February, 2015

Jamail, Dahr, "As Casualties Mount, Scientists Say Global Warming Has Been "Hugely Underestimated", Truthout Report, 20 October 2011.

Jamail, Dahr, "Climate Disruption Depression and 2013 Emissions Set New Records. Truthout, 17 November 2014.

Jamail, Dahr, "Devastating" Impacts of Climate Change Increasing, Truthout, News Analysis, 13 May 2014.

Jamail, Dahr, "Fending For Themselves": Plight of Pointe-au-Chien Indian Tribe in the Wake of the British Petroleum Oil Spill, Truthout Report, 5 July 2010.

Jamail, Dahr, "Former NASA Chief Scientist: 'We're Effectively Taking a Sledgehammer to the Climate System'", Truthout Interview, 30 June 2014.

Jamail, Dahr, "Gulf Ecosystem in Crisis Three Years After BP Spill", Aljazeera English Report, 21 October 2013.

Jamail, Dahr, "Imminent" Collapse of the Antarctic Ice Shelf and a "New Era" in the Arctic, Truthout Report, 1 June 2015

Jamail, Dahr, "Iraqi Doctors Call Depleted Uranium Use 'Genocide'" , Truthout, 14 October 2014.

Jamail, Dahr, "Mass Extinction: It's the End of the World as We Know It", Truthout Interview, 6 July 2015.

Jamail, Dahr, "NASA Scientist Warns of Three to Four-Meter Sea Level Rise by 2200", Truthout Interview, 13 August 2014.

Jamail, Dahr, "Open Source Farming: A Renaissance Man Tackles the Food Crisis", Truthout, Report, 10 August 2014.

Jamail, Dahr, "The Vanishing Arctic Ice Cap", Truthout Report, 31 March 2014.

Jamail, Dahr, "Tortured and Raped by Israel, Persecuted by the United States", Truthout Report, 2 September 2014.

Jamail, Dahr, "Toxic Legacy: Uranium Mining in New Mexico", Truthout Report, 20 February 2014

Jarus, Owen, "The Incas: History of Andean Empire", Live Science Contributor, November 19, 2013.

Jarus, Owen, "The Babylonian Empire", LiveScience Contributor to Ancient Warfare Magazine, ©2010, All Rights Reserved

Jarus, Owen, "The Incas: History of Andean Empire", Live Science Contributor, November 19, 2013

"Why The Colorado River Stopped Flowing", by Staff, National Public Radio, July 14, 2011

Ji, Sayer, "Why The Law Forbids The Medicinal Use of Natural Substances", Green Med Info, February 24, 2015. Greenmedinfo.com

Jibson, Randall W., "Landslide Hazards at La Conchita," California, for United States Geological Survey, 12 January 2013.

Johansen, Bruce E., "The High Cost of Uranium in Navajoland", Greenpeace, Washington D.C., c2014.

Johnson, Lyndon B. (1908-1973) 36th President of the United States, "Transmitting an Assessment of the Nation's Water Resources," Letter to the President of the Senate and to the Speaker of the House 18 Nov 1968

Eric Jungels, "Taking the Plight of Homeless Veterans as His Own," December 22, 2011

Kampwirth, Kevin, "Can Animals Really Anticipate Natural Disasters?" May 20, 2013.

Kane, Alex, "Not Just Ferguson: 11 Eye-Opening Facts About America's Militarized Police Forces", AlterNet, August 13, 2014.

Karlin, Mark, "How the Prison-Industrial Complex Destroys Lives", an interview of Marc Mauer Truthout, Friday, 26 April 2013.

Karunananthan, Meera, "El Salvador Mining Ban Could Establish a Vital Water Security Precedent", Bill and Melinda Gates Foundation, 10 June 2013.

Katch, Danny, "Marked for Death by Trump When he was 15", Social Worker Interview, 28 October 2016.

Katel, Peter, "Do Illegal Workers Help or Hurt the Economy?" a well-documented Abstract, the Pew Hispanic Center, March 21, 2005 with sources, which include: March 2004 "Current Population Survey", by The Census Bureau and Department of Labor.

Kaufman, Alexander C., "Krugman Demolishes Classic Argument Against Raising Minimum Wage", The Huffington Post, 9 September 2014. Paul Krugman is a Nobel Prize-winning economist.

Keefe, Alexa, "Gerd Ludwig's Long Look at the Chernobyl Disaster" interviewing Gerd Ludwig, 30 March 2014.

Kelly, Kathy, "A Teacher in Kabul" (Kathy@vcnu.org), Kathy Kelly, Coordinator, Voices for Creative Nonviolence (www.vcnu.org), August 29, 2014.

Kennis, Andrew, "A Tale of Two Cities: Ciudad Juarez and El Paso", Truthout, 4 May 2012

Keyes, Scott, "Colorado Proves Housing The Homeless Is Cheaper Than Leaving Them On The Streets", September 5, 2013.

Keyes, Scott, "It Costs $21,000 More to Ignore the Homeless Than It Does to Give Them a Home," May 28, 2014.

Keyes, Scott, "The Problem with Criminalizing Homelessness", September 19, 2013

Kiel, Paul, "The Bailout: By The Actual Numbers", ProPublica Journal, September 6, 2012.

Kilauea eruption updates (Jan 2013 - Oct 2014), Natural Resource Report NPS/NRPC/GRD/NRR—2009/163

Kilgore, James, "Jails: Time to Wake Up to Mass Incarceration in Your Neighborhood", Truthout, 4 March 2015.

Kilgore, James, "Time to Stop Criminalizing Immigrants", Truthout Op Ed., 7 August 2014.

Kim, Ye Seul, et al., "Global Warming: Definition", by Ye Seul Kim, Erika Granger, Katie Puckett, Cankutan Hasar, and Leif Francel.

Kochhar, Rakesh, "A Recovery No Better than the Recession" Median Household Income, 2007-2011, Pew Research Center.

Korten, David C., "Getting to the 21st Century": Voluntary Action and the Global Agenda, for the People-Centered Development Forum, 1 January 1990.

Korten, David, "The Elephant in the Room: What Trump, Clinton and Even Stein are Missing." For YES Magazine, October 5, 2016.

Korten, David C., "To Change the Future, Change the Story", Guiding Lights Conference, Seattle 2012. Based on David Korten, *The Great Turning: From Empire to Earth Community.* www.livingeconomies-forum.org

Krauss, Clifford, "Oil Prices Up 3% on News of Economic Growth in Europe, but Supplies Rise" February 13, 2015.

Kraybill, Ken, "Changing the Conversation: The Doctrine of Discovery: A Legacy of Disgrace", Social Justice, January 15, 2015.

Kromm, Chris, "Students Step Up Efforts to Protect Youth Vote", Facing South, News Analysis, 30 August 2014.

Kumar, Hari, "Narendra Modi, favoring growth in India, sweeps away environmental rules", The New York Times, 4 December 2014.

Kunichoff, Yana, "Undocumented Immigrants and Taxation: The Price That We Pay", Truthout, Report, 6 June 2010.

LaDuke, Wanona, "Onaabani-giizis 2015 Newsletter", Written by Wanona LaDuke, Honor the Earth, ©2015.

La Grange, Maria L., "Shell lawsuit against environmental groups ruled unconstitutional", Environmental Issues, Environmental Pollution, Los Angeles Times ©2014.

Lackner, Klaus, "Air Capture and Mineral Sequestration: Tools for Fighting Climate Change", S. Ewing Worzel, Professor of Geophysics, Columbia University, New York.

Lambert, Lisa, "Hold-out creditor argues for dismissal of Detroit bankruptcy case" Reuters, 28 July 2014

Lavelle, Marianne, "D.light attacks a big energy problem with small solar". The Daily Climate, April 21, 2015.

Lavelle, Marianne, "New - and worrisome - contaminants emerge from oil and gas wells", The Daily Climate, Jan. 14, 2014

Lavelle, Marianne, "Solar lights a healthy – and empowering – path in disasters". The Daily Climate, February 26, 2015.

Moira Lavelle, "What's Happened to Brittany, Jonny and Kaylie?" Following up on three poverty-stricken children. July 22, 2014.

Lamb, H. H., "Little Ice Age", Environmental History, "Climatic Fluctuations", World Survey of Climatology. Vol.2. General Climatology (New York: Elsevier, 1969.

Law, Victoria, "If the Risk Is Low, Let Them Go": Efforts to Resolve the Growing Numbers of Aging Behind Bars, Truthout News Analysis, 19 January 2014.

Law, Victoria, "Public Prisons, Private Profits", (About the Prisons Industry, USA), Truthout Report, November 1, 2014.

Lejon, Anna G. C., Renofalt, Birgitta Malm and Nilsson, Christer, "Conflicts Associated with Dam Removal in Sweden", Copyright © 2009 by the author(s). Published here under license by The Resilience Alliance.

Letman, Jon, "A Cynical Environmentalism: Protecting Nature to Prepare for War, Truthout Report, 5 October 2016.

Letman, Jon, "The Militarized Pacific: An Anniversary Without End", Truthout Op-Ed., 14 May 2014.

Levitt, Justin, "A Comprehensive Investigation of voter impersonation finds 31 credible incidents out of 1,000,000,000 ballots cast." The Washington Post, August 6, 2014.

Lewis, Tanya, "Florida Isn't the Only State to 'Ban' Climate Change", Staff Writer for Live Science, March 9, 2015. By

Lichtman, Richard, "Another Descent Into Hell", A Gaza Strip Tragedy, Truthout Op-Ed., 5 August 2014.

Lioz, Adam, "Stacked Deck: How the Bias in Our Big Money Political System Undermines Racial Equity", for Demos, December 11, 2014.

Liptak, Kevin, "U.S. Sending $3 billion to Poor Countries to Combat Climate Change", CNN White House Producer, November 15, 2014

Lipton, Eric, Williams, Brooke and Confessore, Nicholas, "Foreign Powers Buy Influence at Think Tanks", Brookings Institute, September 6, 2014.

LiveScience Staff, "Drought Conditions Worsen in Parts of U.S.", July 8, 2006.

LiveScience Staff, "Second Tropical Cyclone Ever Forms in South Atlantic", March 11, 2010.

Logan, John, "Why Are GOP Politicians and Anti-Union Groups Interfering With the VW Vote?" Truthout Op-Ed., 14 February 2014.

Lopez, Ian Haney, Dog Whistling About ISIS — and Latinos Too, September 30, 2014. His writings have appeared across a range of sources, from the Yale Law Journal to The New York Times.

Lopez, Ian Haney, "How the Politics of Immigration is Driving Mass Deportation" October 7, 2014

Los Angeles Times Staff, "Deadliest U.S. mass shootings - 1984-2015", Oct. 1, 2015.

Loungani, Prakash and Razin, Assaf, "How Beneficial is Foreign Directly Direct Investment for Development Countries"? by Quarterly Magazine of the IMF, Volume38, No. 2, June 2001.

Lubragge, Michael T., "American History, from Revolution to Reconstruction and Beyond", University of Groningen, ©1994-2012 GMW.

Ludwig, Mike, "Carbon Capture and Clean Coal: Obama's Multibillion-Dollar Climate Pipedream", Truthout Report, 3 October 2014

Ludwig, Mike, "Coal Ash: The Dan River Spill and the Power Industry's Toxic Legacy", Truthout Report, 12 February 2014.

Ludwig, Mike, "Food Stamp Outage Highlights Problems With Privatization of Public Services", Truthout, 25 October 2013.

Ludwig, Mike, "Massive Coal Ash Spill Chokes North Carolina River as EPA Considers Waste Rules", Truthout News Analysis, 6 February 2014.

Ludwig, Mike, "Sex Work Wars: Project ROSE, Monica Jones and the Fight for Human Rights", Truthout Report, 13 March 2014.

Ludwig, Mike, "The Mines That Fracking Built", Truthout Report, 2 May 2013.

Ludwig, Mike, "Want the Truth About New York's Human Trafficking Courts? Ask a Sex Worker", Truthout News Analysis, 26 October 2014.

Lund, Jay R., "Sea level rise and Delta subsidence—the demise of sub-sided Delta islands Sea level rise and Delta subsidence—the demise of subsided Delta islands" Professor of Environmental Engineering, University of California, Davis, March 9, 2011.

Lupkin, Sydney, "U.S. Has More Guns – And Gun Deaths – Than Any Other Country, Study Finds", September 19, 2013.

Lurie Julia, "The Supreme Court Just Dealt a Huge Blow to Obama's Climate Plan", Mother Jones, February 9, 2016.

Macaraeg, Sarah, "How the 'Gold Standard' of Police Accountability Fails Civilians by Design", Truthout Report, April 19, 2015

Macie, Edward A. and Hermansen, L. Annie, "Human Influences on Forest Ecosystems", Southern Research Station, May 2003.

Madden, Thomas F. "The Real History of the Crusades", Christianity Today, 5 June 2005.

Madden, Thomas F., "The Truth about the Spanish Inquisition." *Crisis* Magazine, published by Morley Institute, (October 2003).

Magill, Bobby, "US Energy Shakeup Continues, as Solar Capacity Triples," October 20, 2016.

Mahabir, Cynthia, "Heavy Manners and Making Freedom under the People's Revolutionary Government in Grenada, 1979-1983", Department of Afro-American Studies, San Jose State University, ©1993 Academic Press Limited.

Mahabir, Cynthia, "Rape Prosecution, Culture, and Inequality in Postcolonial Grenada", Feminist Studies22, no. 1, Spring 1996, ©1996 by Feminist Studies, Inc.

Mails, Thomas E., "Buffalo, the Life and Spirit of the American Indian", from The Mystic Warriors of the {Plain, by Thomas Mails, ISBN0-7924-5663-7

Marinkovic, Ariel. "Poor safety standards led to Chilean mine disaster", by Also, see link: See also link: http://www.globalpost.com/dispatch/chile/100828/mine-safety

Markus, Bethania Palma, "Borderland Deaths of Migrants Quietly Reach Crisis Numbers", Truthout Report, 27 July 2014.

Marotta, Daniel C and Coute-Marotta, Jennifer, "Colonialism and the Green Economy: The Hidden Side of Carbon Offsets", Truthout, Report, 23 December 2012.

Marotta, Daniel C and Coute-Marotta, Jennifer, "Colonialism and the Green Economy: Villagers Defy Pressure to Forfeit Farms for Carbon-Offset", Truthout, Report, 13 January 2013.

Marash, Clare Smith and Rossman, Julie, "Understanding Fukushima, Part 1: How a Nuclear Plant Works", for World Science Festival, January 30, 2014

Maskin, Michael, "No Safe Place: How Cities Are Making it Illegal to Be Homeless," August 13, 2014.

Masood, Salman, Starved for Energy, Pakistan Braces for a Water Crisis, February 12, 2015.

Massarani, Luisa, "Brazilian Mine Disaster Releases Dangerous Metals", 21 November 2015.

McCoy, Erin L, "West Virginians Raise Alarm as Research Links Coal Mining to Cancer, Birth Defects", Yes! Magazine Report, 1 March 2014.

McDonnell, Tim, "Are Solar-Powered Homes Jacking Up Everyone Else's Electric Bills?" by Climate Desk Associate Producer, March/April 2015 Issue

McDonnell, Tim, "Pentagon: We Could Soon be Fighting Climate Wars", Climate Desk Associate Producer, Oct. 13, 2014.

McDonnell, Tim, "The Town Almost Swallowed by a Coal Mine", If Germany can't kick its coal habit, can anyone? Climate Desk Associate Producer, Department of Defense, ©2013.

McDonnell, Tim, "There's a Horrifying Amount of Plastic in the Ocean. This Chart Shows Who's to Blame". By Tim McDonnell, for Mother Jones Magazine, February 13, 2015.

McDonnell, Tim, "Washington Is Outdoing California and Texas in Renewable Energy". Renewable-energy consumption by state. Climate Desk Associate Producer, Department of Defense, ©2013.

McGirk, Jan, "Growing Coffee, It's Black, no Sugar" Special to MSNBC, August 2002

McGrath, Matt, World's Largest Marine Protected Area Declared in Antarctica, by Matt McGrath, Environmental Correspondent, 28 October 2016.

McKibben, Bill, "A Summer of Extremes Signifies the New Normal", 4 September 2012.

McKibben, Bill, "The Tipping Point", in Business & Innovation, Climate, Energy, Science & Technology, Sustainability, North America. 3 June 2008.

Megna, Michelle, "Highest and lowest car insurance rates by ZIP code", CarInsurance.com January 22, 2015.

Meisel, Duncan, "Zero. That's the New Number", 350.org, September 9, 2016.

Melton, Bruce, "Arctic Warming and Increased Weather Extremes: The National Research Council Speaks", Truthout, News Analysis, 15 July 2014.

Melton, Bruce, "Climate Change 2013: Where We Are Now - Not What You Think", Truthout, News Analysis, 26 December 2013.

Melton, Bruce, "Expect More Warming Than if Nothing Was Done at All", Truthout News Analysis, 27 August 2014.

Meure, MacFarling, "Ice cores and climate change" - British Antarctic Survey, Atmospheric data supplied by the National Oceanic and Atmospheric Administration, ©2006.

MintPress News, "Nestlé Continues Stealing Water During Drought," (from San Bernardino National Forest), March 12, 2015.

Millennium Development Goals, "Culture and Development": Eradicate Extreme Poverty and Hunger, Millennium Development Goals, (MDG), Fund, United Nations Development Programme, © 2013

Moe, Kristin, "Young Climate Marchers: Support From Boomers Makes Us Feel Less Alone", YES! Magazine, Op-Ed, 26 September 2014.

Moench, Brian, "Coal Conspiracy: Stoking Climate Disaster at the BLM", Truthout, News Analysis, 23 May 2015

Monet, Jenni, "Standing Rock and the Militarized Response to Indigenous Movement around the World." October 14, 2016.

Montaigne, Fen, "An Influential Global Voice Warns of Runaway Emissions", recording the interview by global warming economist Fatih Birol, of the International Energy Agency, 11 June 2012.

Montaigne, Fen, "Effects of Global Warming", for National Geographic Magazine, © 1996-2015

Monte Morin, "A cause for pause? Scientists offer reasons for global warming 'hiatus'," Los Angeles Times, 27 February 2015.

Mooney, Chris, "How solar power and electric cars could make suburban living awesome again," December 24, 2013

Morgan, David, "U.S. to delay key health-reform provision to 2015", Reuters, July 2, 2013

Morris, Jim, "Slow-motion disaster for American workers: Lung-damaging silica, other toxic substances kill and sicken tens of thousands each year as regulation falters", Center for Public Integrity, June 29, 2015.

Moskowitz, Peter, "Earthquake spike pushes Oklahoma to consider tighter fracking regulations", The Guardian, 25 June 2015.

Moyers, Bill, "Fighting for Farmworkers", interviews Baldemar Velasquez on the working conditions, July 19, 2013.

Moyers, Bill, "Naomi Klein on Capitalism and Climate Change". Bill Moyers in conversation with Naomi Klein, Moyers & Company, 19 November 2012.

Moyers, Bill, "Seven Key Takeaways From Joseph E. Stiglitz's Tax Plan for Growth and Equality", reported from discussion between Bill Moyers and Joseph E. Stiglitz, author of Reforming Taxation to Promote Growth and Equity, May 30, 2014.

Moyers, Bill, "Welcome to the Plutocracy", a speech by Bill Moyers at Boston University, Howard Zinn Lecture Series, October 29, 2010

Moyers, Bill, and Winship, Michael, "Turn Left on Main Street", a discussion on the Democratic split on Trade agreements with Asia, June 3, 2015.

Municipal Research and Service Center, "Urban Agriculture - Community Gardening," reported by Municipal Research and Service Center of Washington, updated February 2012.

Myers, John Peterson, "Our Stolen Future", a scientific detective story that explores the emerging science of endocrine disruption, Transnational Institute (TNI) of Policy Studies, July 2014.

Nakaruma, David, "Number of unaccompanied children crossing Texas border dropped sharply in July, Obama administration says", The Washington Post, August 7, 2014

Nader, Ralph, "Migrant Workers and America's Harvest of Shame," United Farm Workers of America, Farm Labor Organizing Committee.

Nall, Jeffrey, "Fast-Food Workers Challenge Stereotypes, Globalize Question of Fairness", Truthout News Analysis, 30 August 2014.

Nall, Jeffrey, "One Hundred Ways to Change the Subject: Plutocratic Fallacies in the Service of Fast-Food Exploitation", Truthout News Analysis, 1 September 2014.

Nassar, Wissam, "In Pictures: Gaza Water Crisis Worsens: With 90 percent of the water unfit for human consumption, Palestinians struggle to meet their daily needs in Gaza.", Aljazeera, 12 May 2014.

National Aeronautic and Space Administration, (NASA), "Contaminated Rio Doce Water Flows into the Atlantic," December 3, 2015, http://earthobservatory.nasa.gov

National Aeronautic and Space Administration, "Nyamuragira Volcano, D.R.Kongo (Africa)," Reported by NASA, Earth Observatory, July 18, 2013.

National Aeronautic and Space Administration, "Taking a global perspective on Earth's climate", Reporting on Climate Research Efforts, 2007.

National Employment Law Project, "Low Wage Job Growth": Data taken from: National Employment Law Project, August 2012 Report

National Center for Children in Poverty, "Poverty poses a serious threat to children's brain development." See study: "Poverty and Brain Development in Early Childhood", a study conducted by the National Center for Children in Poverty, New York City, NY, June 1999

National Center for Missing and Exploited Children, "The Super Bowl is the largest human trafficking event in the country", reported by the National Center for Missing and Exploited Children, http://national. deseretnews.com/article/3412/The-Super-Bowl-is-the-largest-human-trafficking-event-in-the-country.html#XPgYDD0gHESMj72t.99 © Copyright 2015, Deseret Digital Media. All Rights Reserved.

National Climatic Data Center, "Tornado Alley", as described by the National Climatic Data Center (NCDC), National Oceanic and Atmospheric Administration, ©2012.

National Geographic, "What Is Global Warming? The Planet is Heating up – and Fast," by National Geographic Magazine

National Geographic Society, "Marine Pollution", Washington D. C. © 1996-2013 National Geographic Society. All rights reserved.

National Institute of Standards and Technology (NIST), Earthquake Loma Prieta California 1989. Report prepared October 18-26, 1989. See also report: "Performance of Structures During the Loma Prieta Earthquake of October 17, 1989".

National Mine Health and Safety Academy, "Technical Information Center and Library, United States Bureau of Mines Collection, (1910 to 1995)". Abolished January 1996.

National Oceanic and Atmospheric Administration, "Mississippi River Flood History 1543-Present", reported by the National Oceanic and

Atmospheric Administration, (NOAA), National Weather Service, last modified: October 12, 2011.

National Park Service, "Top 13 Parks Threatened by Fracking", by National Park Service, August 13, 2013. www.foodandwaterwatch.org

National Parks Conservation Association, "Hoover Dam: A Treasure of Arizona, Nevada and the World," Washington D. C., ©2009.

National Public Radio, "Why The Colorado River Stopped Flowing", by Staff, National Public Radio, July 14, 2011

National Weather Service, "Cloud Classification", prepared by the National Weather Service, June 2010 Review also w-lmk.webmaster@noaa.gov

Navarro, Santiago and Bessi, Renata, "Across Latin America, a Struggle for Communal Land and Indigenous Autonomy", Translated by Miriam Taylor, Truthout, News Analysis, 20 July 2014.

Navarro, Santiago and Bessi, Renata, "Communal Lands: Theater of Operations for the Counterinsurgency", Translated by Miriam Taylor, Truthout, News Analysis, 11 August 2014.

Navarro, Santiago and Bessi, Renata, "Indigenous People Occupy Brazil's Legislature, Protesting Bill's Violation of Land Rights", Translated by Miriam Taylor, Truthout, 25 April 2015.

Navarro, Santiago and Bessi, Renata, "Mexico: Researcher Raises Alert about Environmental Dangers of Wind Farms," translated by Britt Munro and Sarah Farr, Truthout, Wednesday, 17 September 2014,

Navarro, Santiago and Bessi, Renata, "The Partitioning of Brazil's Ocean and Rivers Threatens Small-Scale Fishing Families", Truthout, 07 December 2014.

Natale, Patrick J. "Congress should Act Now to Avoid a Transportation Fiscal Cliff in 2015", @ASCETweets January 17, 2014

Nieto, Sonia, "Thoughts on the 20th Anniversary of Rethinking Schools", Rethinking Schools, Volume 20, no. 3, Spring 2006.

New York City, "Mayor Bloomberg, Deputy Mayor Holloway and Environmental Protection Commissioner Strickland Visit Upstate Construction Site as Critical Repairs to the Delaware Aqueduct Begin." Official Report released by the Mayor's Office of the City of New York, 4 November 2013. Contact: Marc La Vorgna / Jake Goldman (212) 788-2958 Chris Gilbride / Ted Timbers (DEP) (718) 595-6600

New York State Thruway, "About the New Tappan Zee Bridge" Ongoing Reports, 12 May 2015.

New York Times, "Abu Ghraib, 10 Years Later", by the Editorial Board, April 22, 2014

New York Times, "The Plight of American Veterans", Published: November 12, 2007

Newcomb, Steve, "Five Hundred Years of Injustice": The Legacy of Fifteenth Century Religious Prejudice, 1992.

Newitz, Annalee, "The Greatest Mystery of the Inca Empire was its Strange Economy", August 27, 2013

Nichols, Jeremy, "Guardians Spurs Obama Administration to End Coal Industry Breaks, Ensure Mines Pay For Cleanup". August 16, 2016.

Nijhuis, Michelle, "Electric Production" for National Geographic, August 26, 2014

Nimerfroh, Jonathan, "Nearly Frozen Waves Captured On Camera", Nantucket Photographer, February 26, 2015.

Nimmo, Kurt, "Mexicans Stage Drug Raids inside the U.S.", Infowar.com August 27, 2011.

Norris, Floyd, "U.S. Bank Bailout to Rely in Part on Private Money", February 8, 2009.

Norris, Floyd, "U.S. Chose Better Path to Economic Recovery", NYTimes. com May 3, 2012.

Norton, Ben, "Documents Detail US Complicity in Operation Condor Terror Campaign", Truthout, News Analysis, 23 May 2015.

NOVA, "Mystery of the Megavolcano". Original Public Broadcasting Service. Aired September 26, 2006.

O'Donoghue, Amy Joi, "St. George awash with flash floods from violent storms", Deseret News, July 15, 2012.

Ojibwa, "American Indians and European Diseases", December 28, 2009

Ollman, Bertell, "America Beyond Capitalism: A Socialist Stew Prepared for Liberals and Conservatives". Copyright © Bertell Ollman 2004-2015. All rights reserved.

Olson-Raymer, Dr. Gayle, "The Doctrine of Discovery: Pope Nicholas V issued a Papal Bull,1452". (http://ili.nativeweb.org/sdrm art.htlm).

OneAmerica, "An Age of Migration: Globalization and the Root Causes of Migration": Root Causes of Migration – Fact Sheet, © 2014 OneAmerica ®

Onishi, Norimitsu, "Weak Power Grids in Africa Stunt Economies and Fire Up Tempers", for New York Times, July 2, 2015

Ontario Ministry of Labour, "Mining Health, Safety and Prevention Review," Ontario, Canada, April 15, 2015

Oppel, Richard A. Jr. and Wines, Michael, "As Quakes Rattle Oklahoma, Fingers Point to Oil and Gas Industry", April 3, 2015.

Ortiz, Fabiola, "Brazil's 'Dalai Lama of the Rainforest' Faces Death Threats", (Reprint), August 14, 2014.

Oxfam America, "Metal Mining and Sustainable Development in Central America", by Oxfam America Inc., © 2008

Padowski, J. C. and Jawitz, J. W. "Water availability ranking for 225 urban areas in the United States" 2012. Water availability and vulnerability of 225 large cities in the United States, Water Resources Research, 48, W12529, doi:10.1029/2012WR012335.

Pantsios, Anastasia, "What City Will Run Out of Water First?" EcoWatch, September 3, 2014.

Pasma, Chandra, "What is the Basic Income Guarantee?" Basic Income Guarantee Network, June 2014.

Patin, Stanislav, "Accidents During the Offshore Oil and Gas Development", by Stanislav Patin, translated by Elena Ascio, based on "Environmental Impact of the Offshore Oil and Gas Industry", Copyright EcoMonitor Publishing, New York.

Paton, Carol, "New Power Producers Overwhelm 'Weak' Grid", Business Day Live, July 4, 2015.

Paton, Carol, Eskom's Problems Look as if They are Here to Stay for Now", Business Day Live, 1 September 2015.

Patterson, Brittany, "Alaska's 'new normal' dog sled race -- bumping along in the mud is just part of it", E&E reporter, ClimateWire, April 2, 2015

Patterson, Rob, "Natural Gas has Risen to be the Prominent Energy Source of theFuture", Shale Oil and Gas Business Magazine, November 14, 2014.

Paul, Daniel N., "We were not the Savages!" The Hidden History of the Americas, Posted 6 May 2011.

Paul, Daniel N., "American Indian Genocide,", ©2008. www.danieln-paul.com

Paulas, Rick, "'Behind The Kitchen Door' & Fair Wages with KCET!" ROC-U, (Restaurant Opportunity Center, United), March 7, 2013.

Pearson, Catherine, "Canine Telepathy: Can your Dog Read your mind?", Huffington Post, August 12, 2011.

Peeples, Lynne, "Surprise Finding Heightens Concern Over Tiny Bits Of Plastic Polluting Our Oceans", 23 March 2015. lynne.peeples@huffingtonpost.com

Penn, Ivan, "Duke Energy to Cancel Proposed Levy County Nuclear Plant", Times Staff Writer, August 1, 2013.

Pener, Degen, "How 4 Iconic Places in Los Angeles Are Saving Water", National Geographic, May 29, 2015.

Pepitone, Julianne, "Siberian Holes Could Be 'Visible Effect' of Global Warming, Experts Say", NBC News, Science, August 7, 2014.

Perlman, Howard, et, al., "The World's Water" United States Geological Survey, 17 March 2014. URL: http://water.usgs.gov/edu/earthwhere-water.html

Pew Research Center, "The Lost Decade of the Middle Class". Study conducted by the Pew Research Center, with data from the U.S. Census Bureau and Federal Reserve Board of Governors. ©2012.

Pianin, Eric, "Study Finds Illegal Immigrants Pay $11.8B in Taxes", The Fiscal Times, April 16, 2015.

Pilger, John, "Australia Is Again Stealing Its Indigenous Children", Truthout | News Analysis, 25 March 2014.

Pilger, John, "The Forgotten Coup - How the US and Britain Crushed the Government of Their 'Ally' Australia", Truthout, News Analysis, 25 October 2014.

Pitt, William Rivers, "The Iraq War Was a Smashing Success", Truthout Op-Ed., 28 August 2014.

Pitt, William Rivers, "TPP in the USA Is Why We Occupy", Truthout Op-Ed., 21 September 2013 / truth-out.org

Plumer, Brad, "Arctic sea ice hit a record low in 2012. Here's why it matters", August 28, 2012 © 1996-2015 The Washington Post.

Plumer, Brad, "Study: The Gulf Stream system may already be weakening. That's not good." MARCH 23, 2015. brad@vox.com

Pollan, Michael, "Deconstructing Dinner: 'The Omnivore's Dilemma.'" Reviewed by David Kamp. Published: April 23, 2006.

Pollin, Robert, "Build the Green Economy", Boston Review, News Analysis, 18 September 2014.

Pope Francis: "The Earth, our home, is beginning to look like an immense pile of filth", An extract from Pope Francis's encyclical on climate change, the environment and inequality, by Pope Francis, at the Vatican, 18 June 2015.

Porter, Gareth, "The Real Story Behind the Republicans' Iran Letter", Middle East Eve, March 17, 2015.

Public Broadcasting Service, "Comparing International Health Care Systems", PBS NewsHour, October 6, 2009 http://www.pbs.org/newshour/images/primary2/gwen-ifill.jpg

Public Release: 6-Jan-2014: Suburban sprawl cancels carbon footprint savings of dense urban cores: Interactive maps of US metro areas shows striking differences between "Arrested in Ferguson in Act of Repentance" by Jim Wallis, Sojourner Magazine, October 14, 2014.

Quiroga, Javiera, "Bare Slopes Leave Chile Ski Resorts Feeling Like California", June 19, 2015.

RCL Enterprises, Inc., "A Brief History of the Popes", ©2007 RCL Publishing, LLC. All rights reserved

Race, Poverty and the Environment, a journal for social and environmental justice, published by Urban Habitat, volume16, #2, Fall 2009.

RachelCarson.org, "The Life and Legacy of Rachel Carson", Connecticut College, New London, Connecticut, ©1996-2015 RachelCarson.org All Rights Reserved.

Rakia, Raven, "Between the Peacekeepers and the Protesters in Ferguson", Truthout, News Analysis, 9 September 2014.

Randall, Tom, "Fossil Fuels Just Lost the Race Against Renewables", Bloomberg Business News, April 14, 2015.

Rapoport, Miles and Wheary, Jennnifer, "Running in Place: Where The Middle Class and The Poor Meet", New York, NY comm@ demos.org

Reddy, Christopher M. and Valentine, David, "Where Did the BP Deepwater Horizon Oil Go?" Tracking where and how oil traveled in the deep ocean isn't easy, October 27, 2014.

Reed, Stanley, Total, French Oil Giant, Posts $5.7 Billion Quarterly Loss, February 12, 2015.

Reece, Dayne, "Kayaking in the Caroni Bird Sanctuary, introducing a bird sanctuary in a natural wetland, with a variety of abundant aquatic and dry land life," by Dayne Reece, Hikers, Inc. of Trinidad and Tobago, February 22, 2015.

Republican National Committee, "Restoring the American Dream: Economy & Jobs". ©2012-2014, all rights reserved.

Resilience Alliance, "Coastal Lagoons and Climate Change: Ecological and Social Ramifications in U.S. Atlantic and Gulf Coast Ecosystems,"

Ecology and Society, volume 14, No. 1, ©2009. Published here under license by The Resilience Alliance.

Revkin, Andrew C., "What's That Swimming in the Water Supply?; Robot Sub Inspects 45 Miles of a Leaky New York Aqueduct", June 7, 2003

Rex, Erica, "Will judges force the Netherlands to meet E.U. climate goals?" E&E Correspondent, ClimateWire, March 12, 2013.

Ritterman, Jeff, M. D., "Four Decades of the Wrong Dietary Advice Has Paved the Way for the Diabetes Epidemic: Time to Change Course", Truthout News Analysis, 5 June 2015.

Ritterman, Jeff, M. D., "Monsanto's Herbicide Linked to Fatal Kidney Disease Epidemic: Could It Topple the Company?" Truthout News Analysis, 10 July 2014.

Riversong, Robert, "Turning the Tide: Shifting the Paradigm of Human Culture," December 2015, copyleft by Robert Riversong: may be reproduced only with attribution for non-commercial purposes and a link to this page

Robertson, Joshua, Under the sun: "Australia's largest solar farm set to sprout in a Queensland field", 1 March 2015.

Robinson, William I., "The Political Economy of Israeli Apartheid and the Specter of Genocide", Truthout News Analysis, 19 September 2014.

Romain, David, "Essential Neighborhood Ingredients for Lower Income Housing Production and Maintenance", presented to the Third International Symposium on Lower Cost Housing Problems, Montreal Canada, May 27-31, 1974.

Rosencranz, Armin, "U.S. Climate Change Policy under G. W. Bush, GGU Law Digital Commons, 32 Golden Gate U. L. Rev. 2002 http://digitalcommons.law.ggu.edu

Rotondaro, Vinnie, "Doctrine of Discovery: A Scandal in Plain Sight, The Trail of History", September 5, 2015.

Rumberger, Debbie Harrison, "How Gov. Scott Quietly Stole Florida's Future, The Daily Climate, October 24, 2016.

Rumpler, John, "Shalefield Stories Environment America Research and Policy Center", January 30, 2014.

Rusbjerg Jane, Danish Energy Agency: Special Advisor, Centre for Climate and Energy Economics, August 2013.

Rushkoff, Douglas, "Think Occupy Wall St. is a Phase? You Don't Get It". Special to CNN, October 5, 2011

Santich, Kate, "Cost of homelessness in Central Florida? $31K per person", Orlando Sentinel, May 21, 2014.

Sarthou, Cyn, "No convention center was damaged in the BP oil spill" (guest opinion), Executive Director, Gulf Restoration Network, April22, 2015.

Satterthwaite, David and colleagues, "The Ten and a Half Myths that may Distort the urban Policies of Governments and International Agencies", at the Human Settlement Programme, International Institute for Environment and Development, (IIED), 1979.

Sauter, Michael B. et. al., The 13 Worst Recessions, Depressions, and Panics In American History, by Michael B. Sauter, Douglas A.

McIntyre, and Charles B. Stockdale, 247Wall Street Newsletter, September 9, 2010.

Scheinman, Ted, "The Poetry and Politics of Pope Francis' Climate Encyclical", June 16, 2015.

Schiff, Rep. Adam. "The Supreme Court Still Thinks Corporations Are People", Reported by Reuters, July 18, 2012.

Schenwar, Maya, "Selling the Soul of Public Education", Truthout Op-Ed, 06 September 2012.

Schiermeier, Quirin, "Floods: Holding back the tide", (Protecting the Ganges-Brahmaputra delta), 9 April 2014.

Schiffman, Richard, "Sea Change: The Ecological Disaster That Nobody Seas", Trothout, Report, 18 September 2014.

Schjonberg, Mary Frances, "Environmental Stewardship Fellows Foster Ministry Rooted in Creation", January 30, 2015.

Schlanger Zoe, "Man-Made Earthquakes Are Proliferating, but We Won't Admit Fault", by August 25, 2014.

Schori, Katharine Jefferts, "Pastoral letter on the Doctrine of Discovery and Indigenous Peoples," Episcopal Presiding Bishop, The Episcopal Church Office of Public Affairs, May 16, 2012.

Schori, Katharine Jefferts, "Repudiating the Doctrine of Discovery and its impact on Indigenous Peoples": pastoral letter, issued by Bishop Katharine Jefferts Schori, Episcopal Presiding Bishop, The Episcopal Church, Office of Public Affairs, Wednesday, May 16, 2012

Schumaker, Erin, Why Pediatricians are so Alarmed by the lead in Flint's Water, The Huffington Post, December 18, 2015.

Schwartz, John, "Katherine Hayhoe, a Climate Explainer Who Stays Above the Storm, New York Times. Science, October 10, 2016.

Schwartz, Shelly, "Dust Bowl", An Ecological Disaster During the Great Depression, Contributing History Writer, Explore 20th Century History ©2013 About.com. All rights reserved.

Scott, Doug, "ASCE's New [2013]Report Card Bumps the Nation's Infrastructure Grade Up to a D+", American Society of Civil Engineers, ©2013.

Seaman, Greg, "7 Ways Organic Farms Outperform Conventional Farms", October 24, 2011

Seltenrich, Nate, "Between Extremes: Health Effects of Heat and Cold," Petaluma, California, 1 November 2015.

Serrano, Richard A, "Deaths, Defects of Gulf War Vets' Babies Raise Alarm", Los Angeles Times, November 14. 1994

Shahan, Zachary, "13 Charts On Solar Panel Cost & Growth Trends", September 4th, 2014.

Shah, Anup, "Global Issues": Social, Political, Economic and Environmental Issues that affect us all: Poverty Facts and Stats, January 7, 2013. http://www.globalissues.org/article/26/poverty-facts-and-stats

Shah, Anup, "Structural Adjustment—a Major Cause of Poverty", by Anup Shah, a study on global poverty for the United Nations, November 28, 2010.

Sheldrake, Rupert, Listen to the Animals: Why did so many animals escape December's tsunami, BrainHQ Official Site, 30 December 2004.

Sherk, James, "Not Looking for Work: Why Labor Force Participation Has Fallen During the Recovery" by James Sherk, of the Heritage Foundation, published August 2012. Revised and updated September 4, 2014.

Shoko, Janet, "Zimbabwe to Rename Victoria Falls in Anti-Colonial Name Bid," 17 December 2013, Theafricareport.com

Short, Kevin, "Working At McDonald's Is Starkly Different In These 3 Countries", The Huffington Post, 15 May 2014

Simms, Andrew, "Sonderborg: the Little-Known Danish Town with a Zero Carbon Master Plan," 22 October 2015.

Singh, Dylan, "The Motives for 15th Century Spanish and Portuguese Exploration", Lake Ginninderra College 2007

Sitrin, Marina, "Barter Networks: Lessons From Argentina for Greece", teleSur English Op Ed., 19 July 2015.

Slaughter, Anne-Marie, "Occupy Wall Street and the Arab Spring", The Atlantic,October 7, 2011.

Slaughter, Anne-Marie, "The Arab Spring: A Year Of Revolution", National Public Radio staff, December 17, 2011.

Sloane, Amanda, "Teen jailed without conviction sues for $20M", November 26, 2013.

SLUG stands for San Francisco League of Urban Gardeners (San Francisco, CA), July 24, 2003.

Smith, Ellis, "Stroud Watson: The Man Behind Chattanooga's Downtown Revival", Chattanooga Times Free Press, July 31, 2012.

Smith, Heather, "US government says drilling causes earthquakes – what took them so long?" San Francisco, for The Guardian, 24 April 2015.

Smith, Jeff, "Behind the 'Green Economy': Profiting from environmental and climate crisis", Grand Rapids Institute for Information Democracy, September 17, 2012.

Smith, Richard, "Capitalism and the Destruction of Life on Earth", Truthout, Opinion, 10 November 2013.

Smith, Richard, "Green Capitalism: The God That Failed", Truthout News Analysis, 9 January 2014.

Smithsonian, "Environmental Impact of American Indian Farming Prior to European Settlement", Smithsonian Environmental Research Center, 9 May 2011.

Snow, Nick, "Humans Largely Causing Accelerated Climate Change, IPCC Reiterates", Editor, Oil and Gas Journal, Washington D.C., September 27, 2013.

Solomon, Christopher, "When Birds Squawk, Other Species Seem to Listen" for the New York Times, May 18, 2015.

Sosa, Charlotte, "El Salvador Could Become First Country to Ban Metal Mining", for PanAm Post, October 13, 2013.

Sottile, J. P., "We are not Aone: Listening to the 8.7 Million Other Animals Who Live on Earth," Truthout Report, 1 October 2016.

Southwest Florida Water Management District, "Lake drained twice by sinkhole should NOT refill from Florida aquifer". April 16, 2007.

Springer, Hugh W., "Federation in the Caribbean: An Attempt that Failed" for the Harvard Center for International Affairs, ©1962.

Srinivasan, Dr. M., et al. "Parabolic Solar Reflectors", by Solar Cookers International Network. Based on papers presented to a 1978 Solar Energy Symposium.

Stargardter, Gabriel, "U.S. steps up deportation of Central American child migrants," July 18, 2014, Reuters.

Staro, Jim, "So Corporate 1%ers Can't Afford Paying Employee Health Care Insurance", April 14, 2014.

Stein, Jonathan and Dickinson, Tim, Lie by Lie: A Timeline of How We Got Into Iraq by Mother Jones, September/October 2006 Issue.

Steiner, Achim, "Towards a Green Economy, Pathways to Sustainable Development and Poverty Eradication", UN Under-Secretary-General and Executive Director, United Nations Environment Programme, (UNEP), Nairobi, Kenya, 16 November 2011.

Stoll, Mark, "Rachel Carson's Silent Spring, A book that changed the world", New York Times, ©2012.

Storey, William Kelleher, "A Timeline of Modern English History," Resources for Conducting Research and Writing Papers, myweb.fsu.edu

Stromberg, Joseph, "Mining Tar Sand Produces Much More Air Pollution Than We Thought", Science, Smithsonian Magazine, February 3, 2014.

Suisun City Historic Waterfront website: www.suisunwaterfront.com.

Suzuki, David, "Global Warming Deniers Are Growing More Desperate by the Day", David Suzuki Foundation, August 7, 2014.

Swan, Michael, "Doctrine of discovery first repudiated in 1537," The Catholic Register, October 2, 2014.

Taibbi, Matt, "Cruel and Unusual Punishment: The Shame of Three Strikes Laws", Rolling Stone, The Stupidest Law Ever, issue 1180, March 27, 2013.

Tedla, Elleni, "Sankofan Education for Development of Personhood", Center for African American Educational Excellence, California State University, Dominguez Hills, California, ©1997.

Terebay, Jamie, "Obama's Iraq dilemma: Maliki's enemies are not limited to Al-Qaeda", @jamietarabay, January 8, 2014.

Than, Ker, "Deserts Might Grow as Tropics Expand", May 25, 2006.

Thatcher, Wade, "Child Labor During the English Industrial Revolution", c2009

Tierney, John, "Time and Punishment", Mandatory Sentences Face Growing Skepticism, The New York Times, December 11, 2012.

Tihansky, Ann B., "Sinkholes, West-Central Florida", A link between surface water and ground water, by U.S. Geological Survey, Tampa, Florida, ©1987

Todd, Chuck, et.al., "ISIS Speech Offers Obama One Chance to Turn Bad Polls Around", by Chuck Todd, Mark Murray and Carrie Dann, NBC News, September 10, 2014.

Tolan, Sandy, "North Dakota Pipeline Activists Say Arrested Protesters Were Kept in Dog Kennels," Los Angeles Times, October 30, 2016.

Tomasky, Michael, "We Should Negotiate With Terrorists. We Always Have". The Daily Beast, 6 June 2014

Toomey, Diane, "How British Columbia Gained By Putting a Price on Carbon", in interview with Stewart Elgie, Yale Environment 360, 30 April 2015. [which contains 177 reports related to the Yale Environment 360 Forum of April 2015]

Traub, Amy, and Reutschlin, Catherine, "The Racial Wealth Gap: Why Policy Matters", Demos, March 10, 2015.

Tree People, "Institutionalizing Green Infrastructure": The Bright Green Future of Urban Forest Watershed Management, California, ©2003

Trembath Alex, et.al., "Coal Killer, How Natural Gas Fuels the Clean Energy Revolution", by Michael Shellenberger, Ted Nordhaus, Max Luke. Breakthrough Institute, 25 June 2013.

Trickey, Erick, "How Hostile Poll-Watchers Could Hand Pennsylvania to Trump," October 2, 2016.

Troustine, Jean, "Beyond Revenge: Most Crime Victims Prefer Rehabilitation to Harsh Punishment," Truthout Report, 29 October 2016.

Truthout, "Climate in Our Hands: Inside the Ideas and Actions of a Movement", by Truthout Staff, and YES! Magazine, Special Feature, September 2014.

Tuoi Tre News, "Vietnamese Farmers Make Solar-Powered Boat", Dong Thap Province, Vietnam, June 9, 2015.

Turnbull, Lornet, "How Oregon Became the Easiest Place to Vote in America," YES Magazine, September 29, 2016.

Tydfil, Merthyr, "The Disaster at Aberfan", Four Conversations Wales, April 9, 2001.

United Nations, "Doctrine of Discovery, Used for Centuries to Justify Seizure of Indigenous Lands, Subjugate Peoples, Must be Repudiated by United Nations, Permanent Forum Told," by the United Nations Economic and Social Council, 8 May 2012

United Nations Educational, Scientific and Cultural Organization, "Mosi-oa-Tunya/Victoria Falls: Outstanding Universal Value," World Heritage Convention, ©2006

United Nations Framework Convention on Climate Change, Conference of the Parties Twenty-First Session, Paris, 30 November to 11 December 2015.

United States Department of Agriculture, Forest Service, "Urban and Community Forest Technology Transfer", July 2002.

United States Farmworker Factsheet by Student Action with Farmworkers, © 2011 – 2014

United States Geological Survey, "Deadly Lahars from Nevado del Ruiz, Colombia", published by United States Geological Survey, 29 December 2009.

United States Geological Survey, "The Cataclysmic 1991 Eruption of Mount Pinatubo, Philippines," U.S. Geological Survey, Fact Sheet 113-97

University of Illinois, "Cutting Down Trees Doesn't Cut Down on Crime", University of Illinois, Human-Environment Research, Environment & Behavior Volume 33, No. 4, 2001.

University of Michigan, "Human Appropriation of the World's Fresh Water Supply", All materials © 2000 by the University of Michigan.

University of Wisconsin, "Cooperatives in the U.S." Published by The University of Wisconsin Center for Cooperatives. Contact: info@ UWCC.wisc.edu ©2015.

Upton, John, "Massive Louisiana sinkhole just keeps on growing", for Grist. 20 March 2013

Urban, William, "Rethinking the Crusades," Perspectives on History. ©October 1998.

van Gelder, Sarah, "Naomi Klein: People's Climate March is a 'Glimpse of the Movement We Need'", YES! Magazine – Interviewing Naomi Klein, 7 October 2014.

van Roekel, Dennis, NEA President urges President Obama to focus on economy and inequity, We Broke Iraq and We're Still Paying for the Damage by Huffington Post, Posted: December 17, 2013.

Vaughan, Adam, "Keep it in the Ground: Renewable Energy Breaks New Records, Adam Vaughan, Environmental Editor, the Guardian, 25 October 2016.

Vives, Ruben, et.al., "Huge Exxon Mobil explosion a reminder of refinery dangers", by Ruben Vives, Veroniica Roche and Matt Hamilton, Los Angeles Times, February 18, 2015.

Wallis, Jim, "Arrested in Ferguson in Act of Repentance" Sojourner Magazine, October 14, 2014.

Wallis, Jim, "War Is Not the Answer", Sojourner magazine, Sept. 2014

Walraven, Trish, "Company takes action after dental hygienist's discovery", FOX8WEBCENTRAL, September 17, 2014.

Walsh, Bryan, "This Year's Gulf of Mexico Dead Zone Could be the Biggest on Record," @bryanrwalsh June 19, 2013, based on a study by the National Oceanic and Atmospheric Administration.

Ward, Ken Jr., "Where are all the coal ash dams?" (a question prompted by two dams they discovered that were not on their records) November 9, 2009

Weiser, Wendy R. and Opsal, Erik, "States with New Voting Restrictions since 2010 Election", Brennan Center for Justice at New York University School of Law, June 17, 2014.

Warrick, Joby, "Utilities Wage Campaign against Rooftop Solar" The Washington Post, March 7, 2015

Watkins, Chris, et.al., "Amaranth: Modern Prospects for an Ancient Crop," Based on work by Chris Watkins, Curt Beckmann and Eric Blazek, 1 July 2011.

Watson Stephen T., "In restaurant industry, high turnover and intense competition", The Buffalo News, April 21, 2013.

Watts, Anthony, "The Australian Dust Storm as Seen from Space – Dry lake Eyre not Global Warming?" NASA's Earth Observatory, September 23, 2009.

Weinberger, Daniel, "The Causes of Homelessness in America": Poverty and Prejudice: Sicial Security at the Crossroads, for EDGE, Ethics of Development in a Global Environment, July 26, 1999.

Weiser, Kathy, "Buffalo Hunters, Old West Legends", Legends of America, updated April 2015.

Weiser, Leah, "Southern Ocean Sanctuaries Campaign, Global Penguin Conservation", Global Penguin Conservation Topics: International Ocean, Policy Region: Antarctica, Southern Ocean, February 26, 2014. Contact: Leah Weiser, 202-540-6304

Weisman, Jonathan, "A Rebound Takes Root in Michigan, but Voters' Gloom Is Hard to Shake", for New York Times, September 12, 2014.

Wells, Jason, "Drought covers 100% of California for first time in 15 years", Los Angeles Times, © 2014.

White, Peter, "Foreclosuregate Explained: Big Banks on the Brink", T r u t h o u t, 28 October 2010.

Whitehead, John W., "The War on American Veterans, Waged with SWAT Teams. Surveillance and Neglect, May 23, 2014.

Wiencek, Henry, "The Dark Side of Thomas Jefferson", Smithsonian Magazine, October 2012

Wiener, Aaron, "Here's What Happens When a City Isn't Ready for an Explosion in Family Homelessness," March 20, 2014.

Wilkerson, Isabel, "Mike Brown's shooting and Jim Crow lynchings have too much in common. It's time for America to own up", theguardian. com 25 August 2014.

Williams, Delice, "El Salvador Fights to Protect Water from Mining Contamination" 30 June 2013. Source: The Guardian, Mining Watch.

Winship Michael, "On Democracy, What Matters Today", (Washington's Millionaire Boyz Club), January 10, 2014. BillMoyers.com

Winship, Michael, "Watching a Dark Debate from the City of Light," Democracy and Government, Moyers & Company, October 20, 2016.

Wise, Lindsay, "A Drying Shame: With the Ogallala Aquifer in Peril, the Days of Irrigation for Western Kansas Seem Numbered", by The Kansas City Star, July 24, 2015.

Wong, Edward, "Chinese Glacier's Retreat Signals Trouble for Asian Water Supply", December 8, 2015.

Woodard, Stephanie, "The Police Killings no one is Talking About", In These Times, October 17, 2016.

Woodard, Stephanie, "Voices from the Movement for Native Lives," In These Times, October 25, 2016.

World Bank - Millennium Development Goals – © 2011 The World Bank, All Rights Reserved

World Council of Churches, "Statement on the Doctrine of Discovery and its enduring impact on Indigenous Peoples," issued by the World Council of Churches, at their meeting at Bossey, Switzerland, 14-17 February 2012.

Worldwatch Institute, "Solar Power Reaches 100,000 in Rural India," © 2013 Worldwatch Institute, 1400 16th St. NW, Ste. 430, Washington, DC 20036. worldwatch@worldwatch.org

Worldwatch Report, "Solar Cookers in India," A Worldwatch Report, c2012, Washington D.C. 20036, All Rights Reserved

Yale Environment 360 Forum, "Putting a Price on Carbon: An Emissions Cap or a Tax?" Report from the Yale Environment 360 Forum, 7 May 2009.

Yang, Sarah, "Changing How we Farm Can Save Evolutionary Diversity, Study Suggests", Media Relations, University of California, Berkeley, September 11, 2014.

Yee, Blythe, "What's in Your Cup",: On a Study Trip to Guatemala, students sample the socioeconomics of coffee, MBA Class of '09. *A version of this article first appeared in the* Stanford Business Reporter.

Zabludosky, Karla, "Hunting Humans: The Americans Taking Immigration Into Their Own Hands", July 23, 2014, Newsweek.

Zandstra, Matt, "While Flint Drinks Poison, Nestle is Pumping 200 gallons of Fresh Water out of Michigan Every Minute" Sumofus, us@sumofus.org February 9, 2016.

Zezima, Katie, "Banned in Boston: American Indians, but Only for 329 Years", New York Times, November 25, 2004.

Zielinski, Sarah, "The Colorado River Runs Dry", Smithsonian Magazine, October 2010

Zolfaghirfard, Ellie, "Droughts in the Amazon are speeding up climate change: 'Lungs of the planet' are emitting more CO_2 than they capture", Dailymail.com 5 March 2015.

Epilogue:

As we look to the future of the environmental protection movement, we have to prepare ourselves, individually and collectively to the necessary transformation of our modern lifestyle to one centered on the re-establishment of a lifestyle of stewardship of Mother Earth. I say this without apology, hoping that more and more people grow to appreciate the amazing gift of Earth. Moreover, we should truly honor Earth and the indigenous peoples of our global community, who have given most of themselves as a human sacrifice to the preservation of human survival to this point.

The colonial impact of the Doctrine of Discovery continues to persecute indigenous populations, especially when they arrive at the borders of the industrial countries. In many cases, these are countries, whose economies have been built on the lands confiscated by the missionaries of the Doctrine 500 years ago. They should be celebrated as the descendants of those, who laid down their lives for the ill-fated experiment of colonial expansion. We have that mess to clean up.

Yes, a major part, (possibly the first part), of environmental protection must be "cleanup"! The first prize that came with displacing the natives was to occupy their land. Throughout the Americas, the natives had maintained a rich sense of belonging to the land. For thousands of years before the birth of today's modern lifestyle people lived in personalized communities, in which people knew each other and lived in harmony with the land. They lived in harmony with the other lives, which were also living in harmony with the land. When natives name natural wonders, their names reflect what is beautiful and unique about the wonder. Modern names for natural wonders often are given to honor some national hero. Shouldn't we at least honor Earth for creating the beauty that blesses us?

Earth has had its own lifestyle for over four billion years. Human life began on Earth about one million years ago. When one scales Earth's life into one calendar year, human life began during the last few minutes of December 31st. Columbus arrived in the new world during the last four seconds of December 31st.

We know enough, collectively, to get us back to the customs of living in harmony with Earth's lifestyle. We need to be in more honest conversation with our indigenous populations. We need to accept their leadership and be grateful to those, who kept the wisdom of their ancestors. They are our way out of this climate crisis. Let us build consensus communities around the real task of healing our planet. We owe this to our grandchildren!

73269512R00241

Made in the USA
Columbia, SC
09 July 2017